# ORGANIC MEAT PRODUCTION IN THE '90s

Proceedings of a conference held at Reading University
22 September 1989

Edited by
A.T. Chamberlain, J.M. Walsingham
and
B.A. Stark

CHALCOMBE PUBLICATIONS

**First published in Great Britain by**
**Chalcombe Publications**
**Honey Lane, Hurley, Maidenhead, Berks SL6 5LR**



ISBN 0 948617 19 5



Printed in Great Britain by Gwasg Cambria, Aberystwyth

# CONTENTS

# FOREWORD

## by Professor C.R.W. Spedding

Department of Agriculture, University of Reading, Earley Gate, PO Box 236, Reading RG6 2AT

It is only quite recently that organic methods of production have been seriously considered as a major option for land use in the U.K.

Pioneers have been active for a long time, but there has been a widespread feeling that organic farming was, and would remain, a very minor part of the farming scene. Even now the share of the total market is extremely small.

However, in these days of diversification, there are no grounds for regarding minor sectors as of no importance. A great many of our established crop products are in this category as it is.

Furthermore, when new products and methods are being explored, it is very dangerous to write off ventures that are small. All new ventures start this way.

And it is unhelpful to write off new developments on economic grounds before they have had a chance to demonstrate their economic viability, or before they have arrived at anything like their final forms. Cars and aeroplanes would have been—and probably were—dismissed as impractical as well as uneconomic, when they first appeared.

The popular conception of 'organic' farming—that is without 'chemicals'—has been mainly associated with crops. It has been less clear what was meant by 'organic' animal production, and few people appreciate still the importance of, for example, positive animal welfare in the Standards for Organic Production. Nothing, perhaps, illustrates more clearly the gulf between the popular 'no chemicals' concept (a purely negative notion) and the comprehensive, holistic philosophy of the organic producer.

To the producer, rotational use of the land generally requires livestock and the ways in which these animals are treated are quite as important as the treatment of soil and crops.

The Production Standards published by UKROFS are based on

these organic principles, and the UKROFS Board is still grappling with the detailed problems of extending them throughout the entire human food chain, including processing, distribution, and retailing, not only specific products but mixtures, whether home-produced or imported.

It is timely to discuss the particular problems of organic meat, from 'field to fork' (as it has been described) and these Proceedings bring together the knowledge and experience of people involved at all stages in this process.

# PREFACE

Growing consumer demand for organically-produced foods has resulted in greater interest in them amongst farmers, growers and retailers. 1989 has seen the publication of agreed standards by UKROFS (United Kingdom Register of Organic Food Standards) for organic produce.

Many traditionally organic farmers have concentrated their efforts on cereals, fruit and vegetables. The market for meat and meat products is less well-defined, but changing. In order to meet the challenge of the nineties, farmers, wholesalers and retailers will need to know how methods of producing, processing and packaging meat can satisfy the organic standards.

To help them in this task, the Department of Agriculture in the University of Reading brought together a group of experts in the autumn of 1989 under the chairmanship of Professor C.R.W. Spedding. Their observations on the production of organic meat are recorded here.

A.T. Chamberlain
J.M. Walsingham
B.A. Stark

Reading
September 1989

CHAPTER 1

# THE ORGANIC ETHOS AND STANDARDS AS THEY APPLY TO ANIMAL PRODUCTION

**R. Manley**

Cheshire County Council, Backford Hall, Chester CH1 6EA

## SUMMARY

*The UK Register of Organic Food Standards (UKROFS) was created by Food from Britain, with support from MAFF. UKROFS has stated the principles of organic production and it has the specific task of establishing standards which will allow the use of a logo by registered producers and processers of organic food products who achieve these standards.*

*A European Regulation is being prepared for agricultural and horticultural produce, without reference to livestock, and the EEC has shown considerable interest in the UKROFS standards for meat and livestock. The present UK legal system is more attuned to cases where the presence or absence of detectable substances is observed, than to instances where systems and principles have not been followed.*

## UKROFS

The creation of the United Kingdom Register of Organic Food Standards (UKROFS) by Food from Britain, with the support and encouragement of the Ministry of Agriculture, Fisheries and Food (MAFF), brought together a group of experts who, in attempting to draft standards, initially spent many hours debating the basic principles of organic production and processing. That debate resulted in a statement of principles, given in Appendix 1.1.

It must be emphasised that these principles are not a full description of the ethos of organic production; rather, the statement detailed those matters which UKROFS would address in formulating its standards. In the rest of the work of the Producers' Committee of UKROFS, which drafted the meat production standards, those principles were constantly used for reference and guidance.

UKROFS has a very specific purpose which is described in its terms of reference which are given in Appendix 1.2. Its task can be briefly described as being to establish standards which, if complied with, will allow producers and processors to register and use a logo to announce that their products conform with the published standards. It is not the task of UKROFS to market or promote organic products or principles, but in discharging the Board's terms of reference it is probable that standards will be advanced and made more uniform.

## EEC

The European Economic Community (EEC) has, for a long time, been seeking to develop legislation setting out standards for organic produce. Originally a Directive was prepared, which would then be given effect in each member state by the passing of domestic legislation. However, more recently work has progressed on developing a European Regulation which is of direct effect in member states without further action.

The draft Regulation currently addresses only agricultural and horticultural products, without reference to livestock. However, at an early meeting with Community officials, considerable interest was shown in the UKROFS standards for meat and livestock. It may be that the UK's early work in this area will be of considerable influence in Europe.

## CURRENT UK LAW

There is no definition in law of 'organic'. Few of the simple dictionary definitions are thought to be at all relevant by anyone who knows much about organic production. It is probable that no simple definition can exist when what is seeking to be defined is a system or culture rather than something specific or explicit. The fact that none of the organic sector bodies have adopted a single definition of 'organic', even after many attempts, is significant.

The current provisions of food law and trade descriptions law, which require the truthful description and labelling of products, can be applied to organise claims and statements. However, it is far from easy to assure a criminal conviction within a system which traditionally uses statutory definitions, or in their absence applies common, every day meanings to words.

Research indicates only two prosecutions for falsely described organic produce. Both relate to vegetables sold to consumers which, upon analysis, were found to have traces of pesticide present. It is an interesting thought that, as no appeal was made against these convictions, presumably no defence was made that the produce was in fact organic but had been contaminated. No prosecutions in connection with organic meat are known, although several convictions have been obtained where added water, colour, preservatives or hormones have been detected.

It is evident from these comments that our legal system is attuned to evidence which testifies to the presence or absence of detectable substances more readily than to evidence of observation. Enforcement officers are more likely to proceed with an investigation where empirical evidence exists rather than where theory has been corrupted.

## STANDARDS

No part of the UKROFS Standards should be viewed in isolation. Those rules relating to meat and its production are complemented by the rules on feed, grazing, processing and, of course, the organic principles.

An extract from the UKROFS Standards which is directly applicable to meat is in Appendix 1.3. The full Standards may be purchased from:

The Secretary
UKROFS
Food from Britain
301-344 Market Towers
New Covent Garden Market
London SW8 5NQ

Price £15.00

# APPENDIX 1.1

## UNITED KINGDOM REGISTER OF ORGANIC FOOD STANDARDS

### UKROFS standards for organic production

#### Principles of organic production

1. Organic production systems are designed to produce optimum quantities of food of high nutritional quality by using management practices which aim to avoid the use of agro-chemical inputs and which minimise damage to the environment and wildlife.

2. These systems entail the adoption of management practices which underpin and support the principles and aims of organic production. The principles include:—
   ★ Working with natural systems rather than seeking to dominate them.
   ★ The enhancement of biological cycles involving micro-organisms, soil flora and fauna, plants and animals.
   ★ The maintenance and development of soil fertility using the minimum of non-renewable resources.
   ★ The avoidance of pollution.
   ★ Careful attention to animal welfare considerations.
   ★ The protection of the farm environment with regard to wildlife habitat.
   ★ Consideration for the wider social and ecological impact of the farming system.

3. When applied these principles result in production practices whose key characteristics are:—
   ★ The adoption of sound rotations.
   ★ The extensive and rational use of animal manure and vegetable wastes.
   ★ The use of appropriate inputs.
   ★ Appropriate cultivation, weed and pest control techniques.

## APPENDIX 1.2

## UNITED KINGDOM REGISTER OF ORGANIC FOOD STANDARDS

### Terms of reference and rules of procedure for the board

1. A Board shall be established to administer the United Kingdom Register of Organic Food Standards (UKROFS) which will fulfil the following tasks:—
   - i) To set production standards for organic produce tied to a UK organic logo and a code of production practice.
   - ii) To consider applications from organic sector bodies to check that their proposed standards are at least equivalent to the standard set under (i) and are supported by adequate inspection arrangements.
   - iii) To consider applications from individual producers of organic food not linked to a recognised organic sector body, to ensure that their standards conform to the standard set under (i) and are supported by adequate inspection arrangements.
   - iv) To establish a register of individual organic producers whereby registration provides a right to use the logo.
   - v) To communicate to interested parties the standards approved by the Board and to make publicly available the names of those registered.

## APPENDIX 1.3

## UNITED KINGDOM REGISTER OF ORGANIC FOOD STANDARDS

### UKROFS standards for organic production

### Livestock husbandry

1.0 **General**

1.1 All livestock must be handled, housed and transported under conditions which reflect proper care and concern for their welfare at all times and which comply at least with the requirements of all relevant legislation and MAFF Codes of Recommendations for Animal Welfare.

1.2 The breeds and strains of livestock selected must be suitable for raising under local conditions under an organic regime.

1.3 The use of artificial insemination in the breeding programme is permitted.

**2.0 Origin of stock**

Livestock systems should be planned so that the stock is born and raised on an Organic Unit whether or not brought-in from other UKROFS Registered producers.

2.1 Livestock with the exception of pigs and lambs for finishing may, however, be brought-in from non-Registered sources providing the stock has not been treated with chemical substances, except as provided for in paragraphs 7.2, 7.3 and 7.4 below and except for coccidiostats in poultry starter rations, and must come from livestock units:—

i) that are known to the purchaser;
ii) where the veterinary history of the stock purchased is known and recorded;
iii) where the animal welfare provisions detailed in paragraphs 5.1 to 5.6 inclusive and paragraph 7.6 below are observed.

**3.0 Brought-in stock**

3.1 All brought-in stock must be checked for disease and treated as appropriate, which may include a period of quarantine.

3.2 *Calves*

Calves for finishing may be brought-in from:—

a) Dairy herds, no older than 28 days old. The calves should have received colostrum for at least 4 days and subsequently milk or milk replacers without antibiotics or other growth promoters. Calves over 14 days old must have access each day to dry food with sufficient digestible fibre so as not to impair the development of the rumen;
b) Suckler herds, at normal weaning age providing the provisions of paragraph 2.1 above are observed.

Calves brought-in from non-Registered dairy or suckler herds must undergo a conversion period of 26 weeks before the animal qualifies for UKROFS approved status.

3.3 *Dairy animals*

Up to 10% per year of the animals used to provide dairy products may be brought-in from non-Registered sources but they must undergo a conversion period of 12 weeks before the products qualify for UKROFS approved status.

3.4 *Poultry*

a) Poultry for meat production must be brought-in as one day-old chicks.

b) Pullets for egg production may be brought-in from non-Registered sources no older than 10 weeks of age and must undergo a conversion period of 10 weeks before the eggs qualify for UKROFS approved status.

3.5 *Breeding animals*

Animals for breeding may be brought-in from non-Registered sources providing the provisions of paragraph 2.1 above are observed.

4.0 **Livestock management**

The techniques employed in livestock management must be directed towards maintaining the animal in good health. Systems of livestock management which involve routine use of prophylactic drugs are not permitted. The livestock plan should normally be an integral part of the crop rotation and must provide sufficient land:—

a) to prevent overstocking;

b) to allow for rotational or paddock grazing;

c) to allow the sward to recover or be reseeded;

d) to prevent the build-up of parasites.

Since all outdoor pigs and poultry must have access to suitable shelter, movable buildings will usually be required.

5.0 **Welfare and housing**

5.1 Housing and management must be appropriate to the behavioural needs of the animals or birds. All stock must have sufficient room to stand naturally, lie down easily, turn round, groom themselves, assume all natural postures and make all natural movements such as stretching and wing flapping and to walk about freely at least in accordance with MAFF Codes of Recommendations for Animal Welfare.

5.2 All houses in which livestock of any species are confined, for other than very brief periods or during transit, must be well bedded with straw or other appropriate material and the drainage and other aspects of litter management must ensure that all animals have access to dry lying areas.

5.3 Provided that the rest of the floor is solid and well bedded, up to a quarter of the floor area may be slatted.

5.4 Stalls in which animals are confined individually only while eating and individual cubicles, provided that they are large enough to allow the animals to lie naturally and that the animals have free access to them, are permitted but all other individual stalls, tethers and cages are excluded as are all cages in which birds or animals do not have access to littered floors.

5.5 No animal should normally be housed out of sight or sound of others of its own species. Where for any reason that is unavoidable the animal should be housed to allow it the regular sight and sound of human activity or the company of other compatible animals.

5.6 The plans for livestock systems must allow for the livestock, especially breeding cows and sows, to be kept in reasonably stable groups.

6.0 **Livestock diets**

6.1 All livestock systems should be planned to provide one hundred percent of the diet from feedstuffs produced to UKROFS Standards. However, in cases where this is not immediately possible, the following minimum percentage of the dry matter in the daily diet must conform to UKROFS Standards:—

i) non-dairy sheep and goats at least 95%;
ii) beef animals at least 90%;
iii) dairy stock at least 80%;
iv) non-ruminants at least 70%.

The balance may be brought-in from non-Registered sources providing the source, and as appropriate the composition, and the conditions under which the feedstuff was produced are known to the purchaser. Accurate and comprehensive records of all feedstuffs must be maintained and the farming plan must include provisions for a progressive reduction of the percentage of feedstuffs brought-in from non-Registered sources.

6.2 Up to 50% of the dry matter in the organic part of the diet may come from feedstuffs produced on Registered holdings that are in the process of conversion to organic agriculture under UKROFS Standards providing that in the case of:—

i) grassland: no prohibited treatment has been applied in the previous six months;

ii) arable crops: no prohibited treatment has been applied in the six months prior to planting.

6.3 At least sixty percent of the dry matter in all ruminant diets should consist of either fresh green food or unmilled forage produced to UKROFS Standards.

6.4 The diets should be balanced and of good quality and should not have levels of protein and energy or other additions associated with intensive production. Materials which have been subjected to solvent extraction or other processes involving the addition of chemical agents are prohibited.

6.5 Trace element supplements are permitted only where they are necessary to maintain nutrient balance. If such supplements are used, preference should be given to elements in organic combinations.

6.6 Concentrated vitamins and pure amino acids are permitted when necessary to satisfy normal nutrient requirements but not to produce diets of very high nutrient density designed to achieve very early maturity or high levels of production. Preference should be given to suitable natural sources.

6.7 Fats, oils and fatty acids are permitted provided that they are used to balance ingredients of low energy value and not to produce diets of high nutrient density designed to achieve very early maturity or high levels of production.

6.8 Cattle systems should be planned to allow calves to be fed on milk produced to UKROFS Standards until they are eating solid food. Where in emergencies this proves impossible, milk replacers without antibiotics or other growth promoters may be used.

6.9 For all breeding and milk producing livestock the systems should be planned to make the maximum use of grazing. During the winter, when the diet should be based on home produced hay or silage or green food, whenever weather and soil conditions permit the livestock should have access to forage or other grazing. Silage additives are not permitted.

6.10 As far as possible, systems for producing stores or finished poultry and meat animals should also be based on grazing, but animals may be finished in well bedded spacious yards. Pigs and poultry may be finished indoors.

6.11 Pigs should not be weaned at under six weeks of age except in emergencies where milk replacers or suitable dry diets without antibiotics or other growth promoters may be used.

6.12 Whenever possible, pig breeding systems should be planned to allow the sows direct access to soil and growing green food but all sows should have access to an outside run for most of the breeding cycle.

6.13 All egg production systems must be planned to allow the birds regular and free access to fresh growing green food.

7.0 **Animal health**

7.1 Approved producers, and those in the process of conversion, of meat, poultry or dairy products shall market only healthy animals and birds and their products and they shall not withhold any necessary veterinary treatment from any animal or bird.

7.2 The practices employed in the management of livestock must be directed towards maintaining the animals in good health and preventing conditions where conventional veterinary treatments become necessary. Conventional veterinary treatments may, however, be used:

i) where it is necessary to save the life of the animal;
ii) to prevent suffering; or
iii) to treat conditions where no alternative treatment or preventive management practice is available.

7.3 The use of chemical treatments is prohibited except where used in accordance with paragraph 7.2 above or in compliance with Statutory requirements. If organophosphorus compounds are used in compliance with Statutory requirements then the animals must be permanently marked at the time of treatment and:—

i) in the case of meat animals, the animal immediately and irrevocably loses its UKROFS approved status; or

ii) in the case of dairy animals, the animal must undergo the normal conversion period as indicated in paragraph 3.3 above before the products again qualify for UKROFS approved status.

7.4 The use of feeds containing non-food ingredients intended to stimulate growth or production by modifying the gut microflora or the endocrine system is prohibited except where used strictly for therapeutic treatments in accordance with paragraph 7.2 above.

7.5 All medicines must be used in accordance with their UK product licence. Withdrawal times, between administering the medicine and using the products from the animal, shall be at least double those defined by the product licence or the prescribing veterinarian and shall not be less than one week.

7.6 Animals must not be subjected to any surgical or chemical interference which is not designed to improve the animal's own health or wellbeing or that of the group. Castration and de-horning are permitted where it is judged to be necessary for considerations of safety.

7.7 Precise, accurate and up-to-date records of all treated livestock must be kept, clearly identifying the animal concerned. All conventional and other veterinary treatments including the duration, the brand name and manufacturer of any drug used and the withdrawal times must be recorded.

CHAPTER 2

# SUCKLER PRODUCTION UNDER ORGANIC MANAGEMENT

**R. Young**

Kite's Nest Farm, Broadway, Worcestershire WR12 7JT

## SUMMARY

*Kite's Nest farm is a mixed farm comprising 66 hectares (165 acres) of permanent pasture and 88 hectares (220 acres) of arable land. The organically-managed suckler herd comprises 65 cows of various breeds and is the central farm enterprise. The herd is a closed one to reduce the risk of introducing diseases, and both a stock bull and AI are used. The target is for each cow to produce a calf each year, with heifers having a first calving interval of about 14 months. Heifers calve in the spring and the remainder of the herd calves in summer or autumn. Calves are left with the cows until weaned about a month before calving. Stocking rate is low at about 1 cow and calf per hectare (2.5 acres) but cereals are rarely fed. Animals are allowed access to grazing in the winter whenever conditions permit; the routine use of anthelmintics is avoided and antibiotics are administered only when necessary. Animals are slaughtered at a local abattoir and retailed through the farm shop. Products are labelled as organic. Low production costs and good prices for the meat produced mean that despite the low stocking density, the enterprise is financially stable.*

## INTRODUCTION

Kite's Nest farm is a 154 hectare (385 acre), mixed holding on the scarp slope of the Cotswolds, overlooking the Vale of Evesham, which has been managed organically since we bought it in 1980. Although certain aspects of our system might not transfer exactly to other farms, I believe that much of what we do is capable of wider application and much of what we have learned is worthy of dissemination.

The suckler beef system at Kite's Nest is unique to the farm.

There are several reasons for this. The herd of cattle is the central activity of the whole farm enterprise. Cows and calves are kept as naturally as possible, being allowed to graze or seek shelter throughout the year. As a family we believe passionately in stress-free livestock production, and stress-free livestock slaughtering. The particularly extensive aspect of our system is possible because we have 66 hectares (165 acres) of permanent pasture on free-draining banks, in addition to 88 hectares (220 acres) of arable land which is rotated through short-term leys and cereal crops.

## THE SUCKLER HERD

At Kite's Nest we have 65 suckler cows. The herd was founded in 1948 by a great uncle and taken over by me in 1969. My sister joined the family partnership in 1974, at which time we introduced a dozen Lincoln Red crosses, which had been bred out of my father's pedigree herd of Ayrshires. Fifteen years later seven of those cows are still going strong. For historical reasons there are a number of breeds represented within the herd: Hereford, Welsh Black, South Devon, Lincoln Red, Sussex and Beef Shorthorn. Although predominantly British beef breeds are used, we do have a few Charolais crosses.

In order to avoid as much as possible the introduction of disease onto the farm the herd is entirely closed. The last animal bought in was a Hereford cross heifer, obtained at a week of age from Banbury market in 1978. She brought a severe outbreak of scours onto the farm, which affected several other animals and required a long course of antibiotic treatment to effect recovery.

## BREEDING

A home-bred, seven-eighths Welsh Black stock bull is used on most of the cows, but to minimise incestuous matings and to introduce fresh blood lines, artificial insemination is used on a small proportion of cows, and a fresh stock bull is bred from time to time. In 1973 I went on a three day training course on artificial insemination run by AI Breeders Ltd, and I have found the techniques effective and easy to use on cows, although there have sometimes been problems with heifers.

The intention is for each cow to have one calf a year. We have, on average, two sets of twins per year and two fatalities at birth, but we no longer achieve a 365 day calving index because we allow most heifers to have an extra couple of months before going back in calf. This means that they can be managed as part of the herd, without any special feeding, but yet not lose body weight. For this reason we plan for heifers to calve in the spring. The rest of the herd calves in the summer or autumn. A few cows who fail to conceive quickly, or who for one reason or another are not put back with the bull in time are also allowed to calve in the spring, but we find that most of our cows will produce too much milk if they calve between late April and the end of June. This causes a lot of extra work as one or two quarters then have to be milked regularly until the calf is large enough to take all the milk.

Most calves are born outside, although we do have one 'high tech' device you might not expect to find on an organic farm. This is a close circuit television camera above a calving box, which allows us to keep an eye on heifers or any other potentially difficult calvings, especially overnight, from the farmhouse two hundred yards away.

## HERD MANAGEMENT

The cattle are managed on a herd basis. Calves are left on the cows until weaned naturally. Usually this takes place about four weeks before the next calving. Bull calves then stay in the herd with the cow and younger calf for up to another two years. Yearling heifer calves have to be managed separately for a few weeks when the cows run with the bull, but then rejoin the herd. All the best heifers are kept for breeding and over the years we have built up some fairly large family groups, with up to six generations to be seen grazing together, exhibiting direct signs of kinship across up to four generations and indirect signs across all six.

Stocking rate is relatively low, with around 1 hectare (2.5 acres) per cow and calf providing grazing and conserved forage. No cereals are fed except to cows with twins, to the house cow and to a small but expanding herd of free-range pigs.

The cattle graze the permanent grassland during the summer,

with the 'arable grassland' usually being cut once or twice and then grazed into the late autumn. This allows the older pastures to regrow to provide some keep until the end of the year, and occasionally, in a mild winter through into January. During the autumn hay and straw are fed in the fields. Winter feed is based on two types of silage, precision chopped red clover/ryegrass silage made in clamps, and straight lucerne silage made into big bales.

Routine anthelmintics were used until 1981 when they were cut out completely in order to meet Soil Association standards. Despite grave doubts no problems whatsoever have been encountered. We believe this to be due principally to the low stocking rate and the high proportion of adult cattle in the herd at any one time.

Castration is carried out surgically and, providing that this is done during reasonably dry weather, or in winter plenty of clean bedding is provided, it has been found to be perfectly possible to manage without prophylactic use of antibiotics.

Antibiotics are occasionally used, most particularly for cases of foul foot and summer mastitis—both fortunately fairly rare occurrences now. Oral rehydration therapy, although difficult to practise with suckling calves, is used to treat scours if they occur. Kaolin as a binding agent is also useful. It has been found, by experience, better not to use antibiotics in the vast majority of minor calf ailments. Calves running temperatures up to and sometimes over 40°C usually recover fairly rapidly with common sense nursing (this often means no nursing at all), whereas those calves treated in the past with an antibiotic at the first sign of a problem invariably became ill again within two or three weeks. They were sometimes ill a third and fourth time, requiring a range of antibiotic and other treatments and often experiencing a general loss of thriftiness from which they sometimes did not recover.

High priority is given to assessing the needs and well-being of the herd, which is usually kept in four or five groups. The cattle are rarely restricted to just one field and they are allowed to graze during the winter when weather conditions permit. This does have the major disadvantage of poaching certain fields close to the buildings, but has contributed to a continuous improvement in health, high growth rates and a reduction in the amount of medication required.

## THE PRODUCT

All animals are slaughtered by prior arrangement soon after arrival at a local abattoir. Sides of beef are hung for a minimum of 14 days and then retailed and wholesaled through the farm butcher's shop. Slaughtered cows are processed into a range of beefburger products, all marketed with an organic label.

## THE OVERALL SYSTEM

With costs being kept down overall and good premium prices for the meat, the enterprise is financially viable, despite the low stocking density. I believe the system is successful because of its stress-free nature, the natural suckling, the absence of forced weaning, the absence of disease and the low stocking rate. We find that happy animals grow faster, with most of our bull calves gaining around 0.9 kg per day. We are lucky in having a large farm, with several free-draining fields, and in being able to produce virtually all our own livestock feed. We do, however, buy in bedding straw, reserving the straw from our organic cereals for feeding purposes.

How many, if any, of the aspects of our system are prerequisites for successful organic beef production is open to debate, but I know from my own early experiences that less extensive systems and practices which enforce rigid ideals to the detriment of the animals' happiness, inevitably cause more stress and bring with them the problems of intensification and a greater need for medication.

CHAPTER 3

# ORGANIC BEEF PRODUCTION

**B.G.Lowman**
Edinburgh School of Agriculture, West Mains Road,
Edinburgh EH9 3JG

## SUMMARY

*A trial has been carried out to investigate levels of production and problems encountered with cattle reared organically. Week-old calves were allocated randomly to an organic or conventional, 22 month finishing system, including 2 grazing seasons. The aim was to manage both groups to achieve similar rates of liveweight gain, and identical slaughter weights and levels of finish at the same age, by adjusting the level of concentrates fed. During the winter animals were fed either acid additive-treated silage plus conventional barley, or big bale organic silage and organically grown barley. During the initial 3 month calf rearing stage, organic calves grew significantly slower and the difference in liveweight between the groups at first turnout increased to 34 kg at slaughter. Preventive medicine strategies were adopted for both groups, either conventionally or organically based. The carcases were all marketed as normal beef and, although the killing out percentage of the organic group and their level of leanness were higher, income from this group was lower. Consumption of organic barley was higher and its cost was higher than for conventional barley, thus overall costs were greater for the organic animals. The difference in gross margin per head of £68.77 between the groups implies that for the two systems to achieve the same gross margin, a 12.5 per cent premium would be needed for organically-reared animals. A consumer trial to compare the appearance, cooking quality and eating quality of steaks from the two groups of animals indicated that consumers preferred organic steaks in terms of overall eating quality.*

## INTRODUCTION

With increasing consumer awareness of food quality, production methods and the environment, beef farmers are increasingly questioning the viability of organic production systems. While input/output relationships have been developed for organic crops few, if any, guidelines exist for organic beef production.

In order to provide answers to queries from interested beef farmers, it was decided in 1987 to investigate organic beef production on the Edinburgh School of Agriculture farms, with a view to identifying problems and producing solutions. Organic systems of food production were, at that time, defined as 'systems' which largely excluded the use of synthetic 'man made' products such as fertilisers, sprays, feed additives/preservatives and growth regulators. The new UK Register of Organic Food Standards (1989) now defines organic production systems as "systems designed to produce optimum quantities of food of high nutritional quality by using management practices which *aim to avoid the use of agro-chemical inputs* and which minimise damage to the environment and wildlife".

However, in early 1987 there was little agreement between organic organisations in the proportion of agro-chemical products which could be used. As a consequence, the objective of our trial was *to exclude* ***completely*** *the use of synthetic products and to use homoeopathy to replace conventional medicines to maintain animal health,* ***providing*** *animal welfare was not at risk.*

## TRIAL DESIGN

The trial started in March 1987 with the purchase of 32 week-old Hereford × Friesian bull calves. On arrival, 16 calves were allocated at random to an organic production system and the remainder allocated to a normal, conventional system. Both groups were placed on a 22 month finishing system, encompassing 2 grazing seasons.

When the trial started there were no sown organic meadows available for the organic group. As a consequence, the organic group were allocated to organic meadows which had never been fertilised or sprayed in living memory, situated on relatively poor quality land.

At the start of the trial it was decided to manage both groups so as to achieve similar liveweight gains and hence identical slaughter weights and levels of finish at the same age. This would be achieved by feeding, where necessary, more concentrates to the organic group to compensate for the relatively poor quality pasture and forage, to maintain liveweight gains similar to those of the conventional group.

## FEEDS

The conventional calves were reared in individual pens on dried milk powder for the first 6 weeks after arrival. In addition, commercial calf concentrates, containing both feed preservatives and additives, were available *ad libitum.*

Over both summers the conventional calves grazed sown, fertilised pastures which also provided grass silage, preserved with organic acids, for the winter feeding period. This silage was supplemented with conventional barley, grown using inorganic fertilisers and preserved with organic acids, together with refined and artificially produced minerals and vitamins. Throughout their lifetime the conventional animals received a feed additive, at 150 mg/head/day, to improve the efficiency of feed use.

On arrival the organic calves were penned in groups of 4 and fed whole, fresh cow milk via a teat, although unfortunately the cow milk was not from cows kept organically. Their calf concentrate was made from organic barley, fishmeal (as a source of protein), dried seaweed (as a source of minerals) and cod liver oil (as a source of vitamins). During the summer the organic group grazed organic pastures which also supplied the basis of their winter feed in the form of organic big bale silage, made without additives.

## ANIMAL PERFORMANCE

The performance of the 2 groups is shown in Table 3.1. During the initial 3 month calf rearing stage growth rates were significantly depressed for the organic calves, so that at turnout in June they were 21 kg lighter than the conventional group. Although subsequent levels of concentrate feeding were higher for the organic group, this difference in liveweight marginally increased to 34 kg by the time the cattle were slaughtered.

The major problem during the calf rearing stage was the low intake of the organic calf concentrate which was associated with its poor palatability. Four weeks after arrival the concentrate intake for the organic calves was only 120 g/head/day compared with 610 g/head/day for the conventional calves. In order to encourage higher concentrate intakes molassine meal (a non-organic feedstuff containing preservatives) was included in the organic concentrate. Concentrate intake improved but still failed to reach the level of the conventional group (Figure 3.1).

**Table 3.1.** Liveweight of organically- and conventionally-reared cattle at various stages during rearing

| | Liveweight (kg/head) | |
|---|---|---|
| | Organic system | Conventional system |
| Arrival 12/3/87 | 50 | 50 |
| Turnout 4/6/87 | 90 | 111 |
| Housing 3/11/87 | 172 | 199 |
| Turnout 27/4/88 | 281 | 303 |
| Housing 14/10/88 | 433 | 453 |
| Slaughter 18/1/89 | 498 | 532 |

As a consequence of the low growth rates for the organic group, respiratory disorders, particularly pneumonia, became a major problem. One calf was treated immediately with conventional drugs, 4 were treated with and responded to homoeopathic remedies only and the remaining 9 infected calves began treatment using homeopathic remedies but were eventually changed to conventional drugs on veterinary advice. It would appear that the low concentrate intake, and hence poor performance, was a major factor involved in this incidence of respiratory disease and that homoeopathic treatment of sick animals was only partially effective.

Overall, the organic group received 99.4 percent of its total feed intake from organic sources, the exceptions being cow milk and molassine meal used in the calf rearing stage.

**Figure 3.1.** Difference in concentrate intake between conventional and organic calves

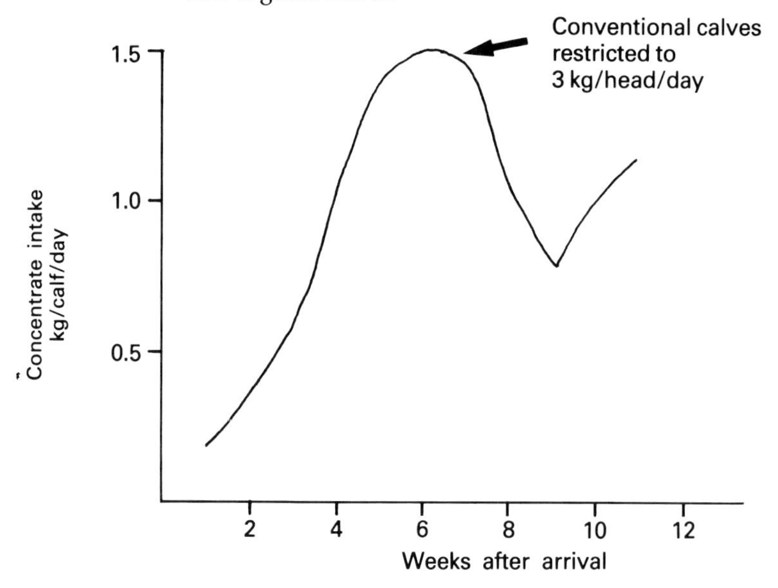


## PREVENTIVE MEDICINE

Homoeopathic vaccines (nosodes) were routinely used in the organic group. On arrival the calves received nosodes to protect against *E. coli* and *Salmonella*, both common causative organisms of diseases leading to severe digestive upsets in young calves. At the start of both summer grazing periods they also received nosodes to prevent lungworm, Foul of the Foot and New Forest Disease (an eye problem associated with flies during the summer grazing period).

Teucrium Marum was administered (1 cc over the throat) monthly over the summer grazing period to control infestations of stomach worms.

In comparison, on arrival the conventional group were given 2 injections of long acting penicillin to prevent the 2 major calfhood diseases, scour and pneumonia. Infections of stomach worms during the 2 grazing seasons were prevented by administering a worming bolus at turnout. At the same time, the calves were tagged with insecticide ear tags to help control flies and hence the incidence of eye diseases.

The conventional calves received a feed additive, to improve the utilisation of feed throughout their life, in the form of a bolus during the summer grazing period and as a supplement in the mineral/vitamin mix during the winter period.

It is important to emphasise that homoeopathic vaccination was used as part of the system and was *not* specifically tested. Hence, although the organic calves did not suffer from any of the diseases for which they had received a nosode, it is possible that the level of challenge to which the calves were exposed was so low that no incidences would have occurred even if the calves had not been vaccinated.

In the first summer grazing period the organic cattle showed symptoms of copper deficiency. Blood samples confirmed this, organic calves having 8.4 μmol Cu/litre compared with 10.7 μmol/litre for the conventional group (normal levels 9.4 to 23.6 μmol/litre). The organic calves therefore received a homoeopathic preparation (Cuprium metallicum) sprayed twice weekly onto their organic barley. Two months later both groups were again blood sampled and the level for the organic group had risen to 13.3 μmol Cu/litre while that for the conventional group was 11.5μmol/litre. The homoeopathic preparation would appear to have improved blood copper levels significantly even though on analysis no trace of copper could be detected in Cuprium metallicum!

A similar problem occurred in trying to evaluate the effectiveness of homoeopathic control of stomach worms. Through the trial both groups of calves were regularly dung sampled and no trace of worm infection was found in either group. This finding is surprising, particularly for the organic calves during the first summer when they were grazed on organic pastures which had been grazed by cattle for the previous 5 years and hence would be expected to have a high level of worm contamination. Turning out young 3 month calves on to such pastures could almost certainly be expected to result in heavy worm infections. Hence, the circumstantial evidence would suggest that Teucrium Marum was apparently effective in controlling levels of worm infection. It must be stressed, however, that whilst the use of homoeopathic preparations appears to have been successful in raising copper status and preventing worm infections, no comparisons with conventional medicines were carried out nor were the potential problems quantified.

## OUTPUT

The output on the 2 systems is presented in Table 3.2. The organic cattle were 34 kg lighter at slaughter but had a slightly better killing out percentage which reduced differences in carcass weight to 15 kg.

The carcasses from the organic group were significantly leaner than the conventional animals and in terms of carcass conformation there was a small, non-significant improvement with the organic cattle.

**Table 3.2.** Output from organically- and conventionally-reared beef cattle

| | Organic cattle | Conventional cattle |
|---|---|---|
| Slaughter weight (kg) | 498 | 532$^{***}$ |
| Carcass weight (kg) | 266 | 281$^{**}$ |
| Killing out % | 53.4 | 52.8$^{*}$ |
| MLC classification | | |
| Fatness | | |
| (3 = 4L; 4 = 4H) | 3.9 | 4.4$^{**}$ |
| Conformation | | |
| (2 = 0+; 3 = R) | 2.2 | 2.0 |

Note: Level of significance $^{*}p<0{\cdot}05$, $^{**}p<0{\cdot}01$, $^{***}p<0.001$

## COSTS

The costs of the 2 systems are presented in Table 3.3. Differences in calf rearing costs at 3 months of age were small, the high price of cow milk for the organic group being partly offset by their much lower intake of concentrates. There were no variable costs associated with grazing and silage making for the organic group, but a total of £68/head was involved for the conventional animals to cover fertiliser costs, reseeding costs and the use of silage additives.

There was a major difference in concentrate intake between the 2 groups, organic calves receiving almost 3 times as much concentrate as the conventional animals in order to achieve similar rates of liveweight gain. Furthermore, the cost per tonne of the organic concentrates, at £176, was considerably higher than the conventional concentrates, due largely to the high cost of purchasing organic barley at £140/tonne.

Over their lifetime, the organic group cost an additional £54.17/head to point of slaughter.

**Table 3.3.** Costs (£/head) for organically- and conventionally-reared cattle

| | Organic cattle | Conventional cattle |
|---|---|---|
| Calf purchase | 140 | 140 |
| Calf rearing | 55.58 | 52.55 |
| Fertiliser/seed/silage additive | — | 68.00 |
| (Silage (tonne/head)) | (3.1) | (6.3) |
| (Concentrates used (kg/head)) | (1025) | (322) |
| Price of concentrates (£/tonne) | (176) | (120) |
| Total concentrate cost | 180.40 | 38.64 |
| Preventive medicines | 8.63 | 31.25 |
| *Total* costs | *384.61* | *330.44* |
| *Difference* in costs | *54.17* | |

## GROSS MARGINS

The gross margins for the 2 groups are presented in Table 3.4. Although the organic group produce carcasses 15 kg lighter than the conventional group, they received a higher average price/kg carcass weight (206 p/kg compared to 200 p/kg). This was solely due to the slightly better conformation and lower fatness for the organic carcasses — all animals being sold as 'normal' beef with no premium being paid for organic status. As a consequence the difference in income was only £14.60 in favour of the conventional animals.

**Table 3.4.** Gross margins (£/head) for organically- and conventionally-reared cattle

| | Organic cattle | | Conventional cattle |
|---|---|---|---|
| Income (£/head) | 549.50 | | 564.10 |
| (Difference) | | (14.60) | |
| Variable costs (£/head) | 384.61 | | 330.44 |
| *Gross margin* (£/head) | *164.89* | | *233.66* |
| *Difference* between groups | | *68.77* | |

The higher production costs and lower income received for the organic group resulted in a £68.77 per head difference in margin between the 2 groups. *For the organic group to achieve the same margin/head as a conventional animal would have required a 12.5 per cent premium to be paid, equivalent to an additional 25 p/kg carcass weight.*

The comparison of economic performance per hectare is complicated by the widely differing quality (and hence rental value) of the land used for the 2 systems. Organic animals required approximately twice the grassland area for grazing and conservation as did the conventional group (1.01 and 0.48 hectare/head respectively). However, had both groups been reared on similar land it is estimated that stocking rates for the organic group would have been approximately 80% of those achieved by conventional cattle. Using this figure with the gross margins per head given in Table 3.4, a 23 per cent premium, equivalent to an additional 48 p/kg carcass weight, would have been required for the organic group to achieve similar gross margins per hectare to the conventional group.

## CONSUMER TRIAL

The question was then asked "Is this level of premium justified in terms of eating quality?"

In January 1989 the 32 cattle were slaughtered. Half of the animals from each treatment (8 animals/treatment) were slaughtered in 2 consecutive weeks. After slaughter 24 sirloin steaks were taken from each animal (a total of 768 steaks) and given to 384 people (2/household). The consumers were asked to cook the steaks in their normal way. Each household received 2 steaks for 2 consecutive weeks. The majority received either an organic steak in the first week and a conventional steak in the second or vice versa. A proportion of the consumers received either organic steaks both weeks or conventional steaks both weeks to act as controls.

*The only information consumers were given was that they would receive 2 sirloin steaks which they were to cook the same way both weeks and that they should fill in the relevant forms.*

### Questionnaires

Each household (2 consumers) received a questionnaire with their steaks:

*Form 1: Appearance*

The first form, to be filled in by the person normally buying meat, related to the appearance of the steaks before cooking.

*Form 2: Cooking*

The person cooking the steaks was asked to fill in a form on how they were cooked, to check that methods were similar both weeks.

*Form 3: Eating quality*

Each consumer was given a form (with their name filled on the top) regarding the eating quality of the steak. In both weeks consumers were asked to compare the steaks with those normally purchased and eaten. In the second week a second group of forms was also distributed, asking consumers to compare the second week's steak with that received in the first week, i.e. a look-back test.

For each question consumers were given a choice of 5 replies:

| | |
|---|---|
| MUCH BETTER | 5 |
| BETTER | 4 |
| THE SAME | 3 |
| WORSE | 2 |
| MUCH WORSE | 1 |

As occurs in all surveys, consumers were reluctant to use extreme markings, i.e. much better or much worse. For the statistical analysis, replies were ranked on a 1 to 5 basis, the higher number indicating the better value. An example of the form is in Appendix 3.1.

## Results of the consumer trial

### *Appearance*

The scores for the appearance of the steaks are presented in Table 3.5. Consumers showed a significant preference for organic steaks in terms of overall appearance (3.44 versus 3.15). However, in looking at the 3 components of overall appearance (texture of lean, colour of lean and leanness) there was only a significant difference for leanness. Organic steaks were classed as being significantly leaner than conventional steaks, reflecting the carcass grading and detailed sample joint dissection.

**Table 3.5.** Appearance of steaks from organically- and conventionally-reared cattle

| | Organic steaks | Conventional steaks | S.E.D. |
|---|---|---|---|
| Overall score | 3.44* | 3.15 | ±0.11 |
| Texture of lean | 3.65 | 3.30 | ±0.19 |
| Colour of lean | 3.46 | 3.23 | ±0.12 |
| Leanness | 3.47** | 2.99 | ±0.16 |

Note: Level of significance *$p<0.05$, **$p<0.01$

*Eating quality*

The scores for eating quality are presented in Table 3.6. In terms of overall eating quality consumers showed a significant preference for organic steaks. The level of statistical significance shows that if there was *no real difference between the 2 steaks* then this result would occur, by chance, once in 100 tests. Whilst differences in flavour, juiciness and tenderness were consistently in favour of organic steaks, the level of significance was less (at the 10% level — occurring by chance once in every 10 tests).

**Table 3.6.** Eating quality of steaks from organically- and conventionally-reared cattle

| | Organic steaks | Conventional steaks | S.E.D. |
|---|---|---|---|
| Overall | 3.44** | 3.24 | ±0.84 |
| Flavour | 3.32 ○ | 3.17 | ±0.85 |
| Juiciness | 3.42 ○ | 3.27 | ±0.81 |
| Tenderness | 3.38 ○ | 3.20 | ±0.94 |

Note: Level of significance ○ $p<0.1$, ** $p<0.01$

## CONCLUSIONS

1. The trial encountered few problems with 99.4 percent organic beef production — suggesting no major obstacles to achieving the current UKROFS standard of supplying at least 90 percent of the total daily dry matter intake from organic feeds. The main problem was the provision of a high quality, palatable organic concentrate for the calf rearing stage. In practice this problem could be overcome by using suckled calves as the basis for an organic beef system.

   There is still, however, no clear list or guide for producers as to what is and is not allowed as an organic feed. For example, is Propcorn treated organic grain acceptable? What about sugar beet pulp? Are silage inoculants allowed?

2. Homoeopathy *appeared* useful for maintaining satisfactory health status in the organic group in terms of preventive medicine using nosodes. However, responses were less effective when treating animals already clinically ill.

3. The £68.77 per head lower margin for organic cattle produced in this trial was largely influenced by the poor quality of the organic pastures available. Organic beef production systems based on high clover, organic swards have been shown to produce similar rates of liveweight gain to fertilised pastures at the North of Scotland College of Agriculture, and this is expected to reduce greatly the differential between the 2 systems.

4. As a consequence, it is likely that the 12.5 percent premium required for organic beef in this trial would be a maximum for the systems to match in terms of gross margins per head.

5. Obtaining a premium for organic beef will require promotion in the market and, most importantly, the guarantee of supplies and consistency of the eating value of the end product.

6. In *this* trial, where organic and conventional cattle were slaughtered at the same age, consumers found a significant preference for organic steaks in terms of overall eating quality. However, there was no one obvious factor influencing overall eating quality, and each characteristic monitored showed a small trend for organic steaks to have a higher value.

# APPENDIX 3.1. LOOK-BACK QUESTIONNAIRE ON THE EATING QUALITY OF ORGANIC AND CONVENTIONAL BEEF STEAKS

TO BE COMPLETED AFTER EATING BY:

________________________________________

________________________________________

ESCA USE ONLY
FORM 3A
DATE
TEST
HOUSEHOLD

SAMPLE

COMPARED with the grilling/frying steak, YOU RECEIVED LAST WEEK what is your opinion of this steak (indicate by ticking one box in each section).

1. OVERALL EATING QUALITY

| | | ESCA |
|---|---|---|
| Much better | ☐ | |
| Better | ☐ | |
| The same | ☐ | ☐ |
| Worse | ☐ | |
| Much worse | ☐ | |

2. FLAVOUR

| | | ESCA |
|---|---|---|
| Much more flavour | ☐ | |
| More flavour | ☐ | |
| The same | ☐ | ☐ |
| Less flavour | ☐ | |
| Much less flavour | ☐ | |

3. JUICINESS

| | | ESCA |
|---|---|---|
| Much more juicy | ☐ | |
| More juicy | ☐ | |
| The same | ☐ | ☐ |
| Less juicy | ☐ | |
| Much less juicy | ☐ | |

4. TENDERNESS

| | | ESCA |
|---|---|---|
| Much more tender | ☐ | |
| More tender | ☐ | ☐ |
| The same | ☐ | |
| Less tender | ☐ | |
| Much less tender | ☐ | |

CHAPTER 4

# ORGANIC SHEEP PRODUCTION

**J.E.Newton**

AFRC Institute for Grassland and Animal Production, North Wyke Research Station, Okehampton, Devon EX20 2SB

## SUMMARY

*The main problem areas of organic sheep production are a low stocking rate leading to a reduced level of income, control of diseases such as clostridia, parasites and fly strike, and effective marketing of the product. This paper presents data from sheep kept at North Wyke Research Station on a no-nitrogen system, and from a commercial farm rearing sheep organically. Stocking rate at North Wyke was 14 ewes and 25 lambs per hectare, with a lamb liveweight gain of 944 kg per hectare; on the farm 12 ewes and 18 lambs were grazed per hectare. Silage was also made from these areas. Yields of fresh herbage and silage were relatively high at both sites, and results from North Wyke suggests that animal performance and overall productivity was much better when animals were grazed rotationally rather than set-stocked. Lamb growth rate under the organic system was satisfactory, with lambs born in April ready for sale in late June; lamb finishing was planned to satisfy market requirements from July to February. Thus organic sheep production need not result in a fall in stocking rate and productivity, whilst disease and parasites can be controlled to a considerable extent by appropriate management.*

## INTRODUCTION

In May 1989 an inaugural meeting of the Organic Sheep Society was held at the Worcester College of Agriculture. Sixty five enthusiastic members attended and the question was raised, very seriously, as to whether there was anyone present who could actually say, with hand on heart, that they had kept to the Soil Association rules and produced organic lamb successfully. It emerged that only a few delegates felt confident that they had succeeded, and there was a widespread awareness of the difficult problems involved. These problem areas, which will be addressed in this paper, are: a decrease in income because of reduced stocking rate; the awful spectre of uncontrollable disease, especially the clostridial diseases; parasites and fly strike; and finally the problem of marketing.

Sheep have now been kept at North Wyke Research Station on a no nitrogen (N) system for 7 years, but I have had no direct experience of keeping sheep organically. For most of the observations that follow I am indebted to Will Best and his wife, who keep 100 Cluns on a 100 hectare organic farm, in a share farming partnership with Mr. and Mrs. Chapman. The farm has a thin chalky, stony soil, and also carries 70 dairy cows together with followers and a few pigs.

## STOCKING RATE

It is assumed that when becoming an organic farm productivity per hectare drops because inorganic fertiliser nitrogen cannot be used. However, this is not necessarily the case, as shown in Table 4.1. Both Will Best's Cluns, and Mashams at North Wyke were grazed on grass/white clover pastures. At North Wyke the stocking rate was held at 14 ewes plus lambs per hectare from mid March to late October; in Will's case the ewes stayed on the area from April to the autumn. His lambs were weaned onto a different area in late July, but dairy heifers were then brought onto the sheep area, thus complicating the stocking rate calculation.

**Table 4.1.** Stocking rate and lamb output

| | Nitrogen (kg/ha) | Ewes + lambs (per ha) | Lamb liveweight gain (kg/ha) |
|---|---|---|---|
| Will Best (organic sheep) | 0 | 12 + 18 | |
| North Wyke | 0* | 14 + 25 | 944 |
| MLC Spring lambing lowland sheep flocks | 167 | 12 + 18 | |

* Phosphate (P) and potassium (K) fertilizer used

## SILAGE YIELD

On Will Best's farm first cut silage was made from half the grass/clover area allocated to the sheep. The dry matter yields, shown in Table 4.2, compare reasonably well with those of a first cut grass sward receiving 120 kg N per hectare. At North Wyke, stocking at 14 ewes and 25 lambs per hectare allowed 90 per cent of the ewes' winter requirements for silage to be made, based on the ewes being housed for 100 days. Additionally, the making of silage from grass/clover with no N fertiliser has been shown to boost the subsequent clover content (Curll and Wilkins, 1985).

**Table 4.2.** Silage yield at first cut (t DM/ha)

| | Nitrogen (kg/ha) | Silage yield (t DM/ha) |
|---|---|---|
| Will Best (grass + clover) | 0 | 4.5 |
| Grass at North Wyke* | 120 | 5.5 |

* From Peel, Matkin, Huckle and Doyle (1986)

## HERBAGE YIELD

The annual yield of grass/clover swards has been shown to depend on the proportion of clover, with yields of 11 tonnes of dry matter (DM) per hectare, when clover was at 45 per cent (see Table 4.3, based on Curll, Wilkins, Snaydon and Shanmugalingham, 1985). This annual yield of 11 tonnes DM per hectare in Berkshire compares well with the 11.8 tonnes DM per hectare with 44 per cent clover measured by Reid (1983) in Scotland. These yields from grass/clover with no N are slightly above the average measured yield of 10 tonnes DM for grass receiving 200kg N (Morrison, Jackson and Sparrow, 1980).

**Table 4.3.** Effect of clover proportion on total yield of herbage (t DM/ha)

| % clover | Yield (t DM/ha) Clover | Grass | Total |
|---|---|---|---|
| 10 | 0.6 | 5.4 | 6.0 |
| 20 | 1.6 | 6.4 | 8.0 |
| 30 | 2.9 | 6.8 | 9.7 |
| 40 | 4.9 | 6.1 | 11.0 |

From: Curll *et al.* (1985)

## GRAZING MANAGEMENT

The advantages of grazing rotationally compared to either set-stocking or continuous grazing, particularly using a 6 paddock rotation, are shown in Table 4.4. Rotational grazing was strikingly better than set-stocking for animal performance and for overall productivity. When compared with continuous grazing in a later experiment, there was still a significant advantage in terms of lamb growth, but the main difference was in overall productivity, strongly suggesting that more herbage dry matter is grown and utilised under rotational grazing.

## LAMB GROWTH

Lamb growth was satisfactory in Will Best's organic system, as it should have been using grass/white clover. Lambs born in April were ready for sale in late June. However, it suited his particular marketing system to have lambs available for sale from July to February, thus rapid growth rates were not required for all the lambs. Customers who bought the handy-sized half carcasses in July wanted to buy more organic lamb from him later in the year.

## GENERAL MANAGEMENT

As the soil is freely draining, the in-lamb ewes were not housed during the winter except for a day or two just before lambing. Concentrates were fed at 600g per head per day at a flat rate for 6 weeks pre-lambing. The concentrate mixture consisted of 3 parts homegrown oats and 1 part of a commercial high protein mix.

**Table 4.4.** The effect of grazing management on the overall production from grass/white clover swards*

| Treatment | Lamb growth rate from day 0 to sale (g/hd/day) | Silage made (kg DM/ewe) | Supplement fed (t DM/ha) | Productivity (GJ/ha) |
|---|---|---|---|---|
| Rotational grazing (6 paddocks) | 225 | — | 0 | — |
| Set-stocking | 142 | — | 1.4 | — |
| Rotational grazing (6 paddocks) | 236 | 46 | 0 | 110 |
| Continuous grazing | 209 | 7 | 1.4 | 98 |

* All treatments at 14 ewes and 25 lambs per ha. Data from Newton and Laws, North Wyke.

## THE GOLDEN HOOF

The sheep were considered by Will Best to make 'a phenomenal contribution' to the overall fertility of the soil, particularly to subsequent cereal yields. The dry ewes were also scavengers, cleaning up the pastures after the dairy cows. Furthermore, the sheep flock also reduced the population of annual weeds in the new leys.

## DISEASE

### Clostridial diseases

Will Best's sheep are not vaccinated against any clostridial diseases and there has been no problem, so far. Before the sheep were introduced 5 years ago there had been no sheep on the farm for 30 years. If pulpy kidney or another soil-borne clostridial disease did occur then the sheep would be vaccinated.

### Flystrike

Some sheep suffered from fly strike in 1988 and were successfully treated with iodine. There are now several pour-on, non-organo-phosphorus compounds available against fly strike, and the Organic Sheep Society is currently conducting a trial with Cypor, the product from Border Research. The results will be available at a later date.

### Other disease problems

Fifteen of Will Best's lambs were lost due to *Pasteurella* in 1988. Too many small lambs were born that year and ewes milked poorly. This year the level of nutrition during pregnancy was increased.

## PARASITES

The fundamental basis of Will Best's farming policy is that there should be a diversity of enterprises. On the livestock side he keeps dairy cows, sheep and a few pigs. He grows oats in addition to wheat, and his grass mixture contains perennial ryegrass, white clover, chicory and plantains.

This enables him, with careful planning, to practise a clean grazing policy. The sheep graze grass/clover that was used for silage and which had no stock at all in the previous year. The lambs are weaned onto the aftermath of second cut silage and then graze the newly seeded pastures and stubble turnips.

He expects to have much less disease on an organic farm, not because the stocking rate is lower, but because intensive 'chemical' farming causes its own problems. He claims that with organic farming the balance of nutrients and minerals in the soil is more likely to be right, although there is as yet no precise definition of the meaning of 'right'.

Alternative methods of reducing the risk of parasites are:

1) To stock with small ewes and singles at a higher stocking rate, single lambs being less dependent on grass than twins or triplets.
2) To lamb early and house the lambs.

## MARKETING

In a recent paper to the Organic Sheep Society, Peter Mitchell listed 4 possible outlets for organic meat. These are:

1) Farm gate
2) Organic butcher
3) Conventional butcher
4) Supermarket

With the conventional butcher and the supermarket there is the problem of organic and non-organic lamb appearing for sale side by side. Despite being more expensive, the organic lamb already outsells the non-organic lamb (New Farmer and Grower, Issue 21, Winter 1988). The question is raised in the mind of the shopper as to what is wrong with the non-organic lamb that it should be cheaper. There is still the problem of educating people to understand the difference between additive-free meat and organic meat, with additive-free meat being cheaper.

## CONCLUSIONS

In summary, there are four heartening messages emerging with regard to organic sheep production:

1) There need be no fall in stocking rate and productivity, since low productivity of grass swards can be circumvented by including white clover and using rotational grazing.
2) Disease can be controlled to a considerable extent by avoiding specialisation and monoculture.
3) Careful planning coupled with grazing by different types of livestock will ensure clean grazing.
4) An unfulfilled demand for organic lamb already exists.

Indeed, the problem with organic lamb is supply not demand,

which at the moment is not being met. This must be a very healthy position for anyone starting organic sheep production.

## REFERENCES

Curll, M.L. and Wilkins, R.J. (1985) The effect of cutting for conservation on a grazed perennial ryegrass-white clover pasture. *Grass and Forage Science* **40**, 19-30.

Curll, M.L., Wilkins, R.J., Snaydon, R.W and Shanmugalingham, V.S (1985) The effects of stocking rate and nitrogen fertiliser on a perennial ryegrass-white clover sward. 1. Sward and sheep performance. *Grass and Forage Science* **40**, 129-140.

Morrison, J., Jackson, M.V. and Sparrow, P.E. (1980) The response of perennial ryegrass to fertiliser nitrogen in relation to climate and soil. *Grassland Research Institute, Technical Report* **27**, 90pp.

Peel, S., Matkin, E.A., Huckle, C.A. and Doyle, C.J. (1986) Can grass growth predictions be used with confidence on the farm? *British Grassland Society, Winter Meeting 1986, Grassland Planning.* 5.1-5.5.

Reid, D. (1983) The combined use of fertiliser nitrogen and white clover as nitrogen sources for herbage growth. *Journal of Agricultural Science, Cambridge* **100**, 613-623.

CHAPTER 5

# EIGHTEEN-MONTH BEEF PRODUCTION: ORGANIC AND INTENSIVE SYSTEMS COMPARED

**D. Younie**

North of Scotland College of Agriculture, School of Agriculture Building, 581 King Street, Aberdeen AB9 1UD

## SUMMARY

*Two self-contained farmlets have been operated at the North of Scotland College since 1983. As far as possible within the constraints of each approach, management of the two farmlets is similar. The intensive system is based on grassland receiving a relatively high level of nitrogen fertiliser; the organic sward is a grass/clover mixture receiving no artificial fertiliser. Overall stocking rates for animals purchased in the autumn at 2 to 3 weeks old up to slaughter at 470 kg average liveweight were 3.6 and 4.3 animals per hectare for the organic and intensive systems respectively. Organic animals received no routine veterinary treatment and, when necessary, homeopathic treatments were preferred; an anti-bloat, molasses based lick was on offer. Various criteria were used to compare output from the two systems including silage yields, average grazing stocking weight per hectare and daily liveweight gains. Overall most parameters were slightly lower for the organic system; daily liveweight gains were higher for the organic group during the two winter periods, but lower during the grazing season, probably reflecting the benefits of clover silage. The gross margin per head was 12 per cent lower for the organic system and 27 per cent lower per hectare, indicating a need for a premium of 30 pence per kg dead weight to achieve the same gross margin per hectare as for the intensive group. The two major technical difficulties in developing a sustainable, long-term system are the difficulty of maintaining the potassium status of the soil and controlling parasitic gastroenteritis.*

## INTRODUCTION

The organic farmer has objectives which go beyond the purely financial, encompassing in addition environmental and animal welfare considerations. Nevertheless his business requires to be profitable if he is to survive, and so the level of physical and financial output per unit area is still of prime concern.

In organic livestock production, output per hectare is governed by the level of grassland productivity. This in turn is dependent, for any given soil type, on the proportion and vigour of the clover component in the sward. Clover was traditionally the mainstay of British agriculture before mineral nitrogen (N) fertilisers became cheaply available half a century ago. Its N-fixing ability provides a source of the nutrient which is required in greatest quantity by the sward (Frame and Newbould, 1986).

Apart from its agronomic effect, clover also has beneficial effects on animal nutrition and performance which have important implications for the organic livestock farmer (Thomson, 1984).

Since 1983 we have operated at Aberdeen an experimental unit which is composed of two self-contained farmlets providing all grazing and winter forage provision for two 18-month beef enterprises. One of these is managed intensively with high levels of N fertiliser input (I); the second has been managed organically (O), according to Soil Association Standards, since 1986. This paper describes the differences in physical and financial performance between these two systems, particularly in terms of the level of output per hectare.

## THE SYSTEMS

Valid comparisons between whole systems are sometimes difficult to achieve because of confounding effects introduced by the variations in management inputs required in each system. In our project both livestock and grassland management regimes were kept identical as far as possible, from choice of grass and clover variety through to grazing management regime.

### Grassland management

The farmlets were located on adjacent sites with similar aspects, topography and sandy loam soils. They were established in May 1983 by direct sowing (i.e. without a cover crop). The seed mixtures were as follows:

| | O | I |
|---|---|---|
| | kg/ha | |
| Bastion early perennial ryegrass | 5.6 | 6 |
| Talbot intermediate perennial ryegrass | 5.6 | 6 |
| Meltra late perennial ryegrass | 7.5 | 8 |
| Preference late perennial ryegrass | 5.6 | 6 |
| Erecta Timothy | 3.7 | 4 |
| Milkanova white clover | 4.0 | 2 |
| | 32 | 32 |

In system O grass seed was drilled and clover seed broadcast; no fertiliser N or herbicide was applied. In subsequent seasons some fertiliser N was applied to this system, until 1986 when all applications of soluble fertiliser ceased (Younie, Heath and Halliday, 1988). Since 1986 only lime has been applied, in addition to approximately 18 t/ha farmyard manure applied to each silage cut. In system I total fertiliser N input was 270 kg N/ha/annum, together with 75 kg $P_2O_5$/ha and 150 kg $K_2O$/ha, and 18 t/ha cattle slurry split between first and second cuts.

A 1:2:3 system of grassland management was followed on both farmlets, in which initially one third of the total area was grazed and two thirds was taken for first cut silage. Animals were transferred from the initial grazed area to first cut silage aftermath in late June, at a reduced stocking rate, and in August stocking rate was reduced again when they were re-admitted to the initially grazed area after it had produced the second silage cut.

Within each grazing area, a buffer fence was moved weekly in order to ensure equal herbage availability, as measured by grass height, to both groups of animals. Target grass height was 7 cm in May-June, rising to 8 cm in July-October. Surplus herbage behind the buffer fence was conserved as big bale silage. Thus, although stocking rates differed between the two systems (see later), any differences in total herbage output between the systems was reflected fairly precisely in terms of surplus silage production.

Molasses was used as an additive on all silage from O, while formic acid was used on the silage from I.

### Livestock management

Autumn-born Hereford × Friesian steers were purchased at 2 to 3 weeks of age, artificially reared indoors, turned out to grass in early May at about 200 kg liveweight (LW), housed again in mid-October at approximately 350 kg LW, and slaughtered between December and March at 17 to 18 months of age at about 470 kg average liveweight. There were 21 animals in each of the two systems, giving overall stocking rates (including the silage areas) of 3.6 and 4.3 animals per hectare for O and I respectively.

### Diet

Organic calves were fed on lamb or kid milk replacer (free of additives) until nine weeks of age, as required by Soil Association Standards, and were offered a home mixed calf concentrate comprised as follows:

| | % by weight |
|---|---|
| Organic oats | 72 |
| Fishmeal | 10 |
| Grass meal | 10 |
| Seaweed meal | 3 |
| Molassine meal | 5 |
| | 100 |

The rearing diet after weaning comprised grass/clover silage produced from the organic farmlet and 2 kg (reducing to 1 kg) of organic oats per day plus seaweed meal. The diet during the fattening phase comprised grass/clover silage *ad lib* plus 2 kg per day of organic oats.

Calves in the conventional group were fed on a similar basic diet but they were given conventional milk replacer to only six weeks of age, along with proprietary calf concentrate, after which they were weaned on to a diet of silage produced from the intensive farmlet plus 2 kg feed barley per day plus mineral mix. The diet in the fattening phase was grass silage plus 2 kg per day of feed barley.

### Veterinary treatment

Organic animals received no routine veterinary treatments. Homeopathic treatments were used where possible.

While the conventional group of animals received injections of Ivermectin at 3, 8 and 13 weeks after turn-out to control parasitic gastro-enteritis (PGE), worm control in the organic group was based on the alternating cutting and grazing management system, with animals being moved on to relatively clean pasture, at lower stocking rates, as the season progressed. Animals which displayed symptoms of PGE late in the grazing season were administered homeopathic treatments or oxfendazole (product Systemex).

The risk of bloat is high on clover-rich pastures, especially on silage aftermaths, and so the organic group of cattle had free access in the second half of the season to a molasses based lick containing poloxalene, an anti-foaming agent.

## PHYSICAL PERFORMANCE

### Output per hectare

Several criteria can be used to compare the physical output per hectare of the systems. First cut silage yields are presented in Table 5.1. On average the organic system produced yields which were 11 percent lower than the intensive system, although still comparatively high in relation to many conventional farm situations.

**Table 5.1.** First cut silage yields from organic and intensive systems (tonnes dry matter per hectare)

| Year | System | |
|---|---|---|
| | O | I |
| 1987 | 6.18 | 6.38 |
| 1988 | 6.01 | 6.73 |
| 1989 | 5.64 | 6.88 |
| Mean | 5.94 | 6.66 |

These relatively high silage yields result in the removal of considerable quantities of potassium from the soil and, despite the return of animal manures to the silage areas, the soil potassium status appears to be declining (Table 5.2).

**Table 5.2.** Changes in soil potassium status of organic farmlet over time

| Sampling date | Extractable soil potassium (g/kg air dry soil) |
|---|---|
| April 1987 | 88 |
| November 1987 | 59 |
| January 1989 | 52 |

Average monthly stocking weight carried per hectare, meaned over the 1987 and 1988 grazing seasons, is shown in Figure 5.1.

Figure 5.1. Average grazing stocking weight per hectare (means of 1987 and 1988 seasons)

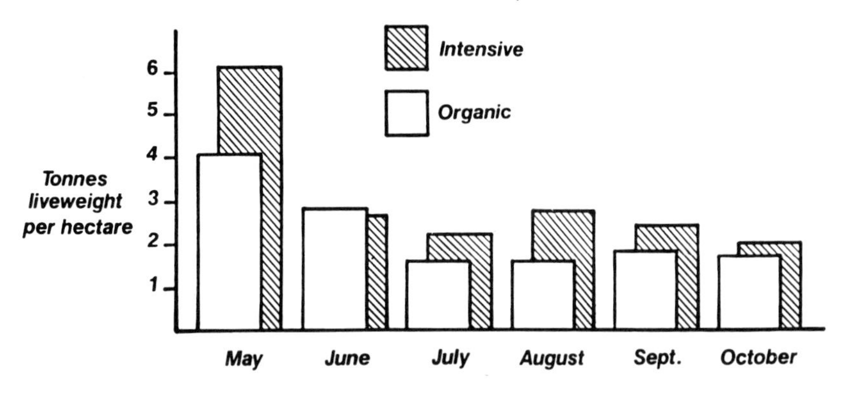


High stocking rates were necessary in the first month after turnout, in order to reduce grass height quickly to the target level. Herbage growth was approximately 10 days later on the organic sward than on the fertilised intensive sward, but turnout date in north-east Scotland is determined much less by target grass height than by climatic conditions. In fact, by the end of May there was usually still sufficient grass buffer remaining on both systems to be cut for silage.

Average annual stocking weights per hectare (mean of six months grazing season) were 2250 and 2994 kg LW/ha for O and I respectively, a difference of 25 per cent, but both figures are high relative to Meat and Livestock Commission (MLC) 18-month beef recorded herds (MLC, 1988).

Liveweight gains (LWG) per hectare over the whole 1987-89 production cycle were 1469 and 1797 kg/hectare for O and I respectively, i.e. a difference of 18 per cent.

## Individual animal performance

Individual animal performance obtained in the 1987-1989 production cycle is presented in Table 5.3.

**Table 5.3** Individual animal performance from organic and intensive 18-month beef systems (1987-89 production cycle)

| | O | I |
|---|---|---|
| Starting weight (kg) | 60 | 60 |
| Daily liveweight gain (DLWG) (kg/head) | 0.77 | 0.67 |
| Turnout weight (kg) | 221 | 195 |
| DLWG (kg/head) | 0.68 | 0.92 |
| Housing weight (kg) | 339 | 368 |
| DLWG (kg/head) | 1.23 | 1.03 |
| Slaughter weight (kg) | 472 | 475 |

The organic group performed significantly better during the rearing and fattening winters, but significantly poorer during the grazing season, compared with the intensive group. The relatively poor performance at grass was caused by a combination of the compensatory growth phenomenon following good performance in the rearing winter, coupled with the effects of stomach worm infection late in the season. Performance recovered well during the subsequent fattening winter.

The good performance of the organic animals during both housing phases is a reflection of the nutritional benefits of clover silage. This factor offers an opportunity to reduce the input of costly organic cereal in the diet, whilst maintaining individual performance. The performance of the most recent cycle of animals illustrates this (Table 5.4).

**Table 5.4.** Liveweight gain from organic and intensive animals during the rearing winter (January to May 1989)

| | O | I |
|---|---|---|
| Concentrate (kg/hd/day) | 1.0 | 2.0 |
| Daily liveweight gain (kg/hd) | 0.70 | 0.60 |
| Turnout weight (kg) | 213 | 196 |

Thus, although given a concentrate allowance of only half that of the conventional animals, the organic group gained weight 17 percent faster and were turned out 9 percent heavier. The considerable nutritional potential of clover-based systems has also been illustrated by Peel, Mayne, Tichen and Huckle (1988).

However, the extent to which cereal inputs can be reduced is limited by the availability of silage. Well made clover silage tends to be more palatable than grass silage, and reducing cereal input increases daily intake even further. Table 5.5 shows the silage requirements of the two systems at Aberdeen. In addition, the lower levels of silage yield relative to the intensive system must be taken into account, but our experience of this self-contained unit has shown that the present stocking level can be supported in terms of winter forage supply, provided concentrate input in the fattening winter is maintained at around 2 kg per head per day.

## Parasitic gastro-enteritis

It was hoped that stomach worm infection could be minimised in this continuous beef situation by the alternating grazing and cutting management, coupled with the relatively short summer season in north-east Scotland which generally allows only one complete cycle of worms to develop per annum.

Biochemical analysis of blood samples taken regularly throughout the grazing season was used to monitor the build-up of infection. Blood pepsinogen content is a reliable indicator of gut wall damage by *Ostertagia*. In both the 1987 and 1988 seasons blood pepsinogen contents of the organic group were elevated at an earlier stage in the season, and rose to higher levels overall than the conventional group (Figure 5.2).

**Table 5.5.** Silage requirements for organic and intensive 18-month beef systems

| | Fresh silage (t/head) | | Dry matter (kg/head) | |
|---|---|---|---|---|
| | O | I | O | I |
| First winter | 2.03 | 1.45 | 407 | 267 |
| Second winter | 3.21 | 3.22 | 721 | 664 |
| Total | 5.24 | 4.67 | 1128 | 931 |

Figure 5.2. Blood pepsinogen levels in animals untreated or treated with ivomectin, 1987

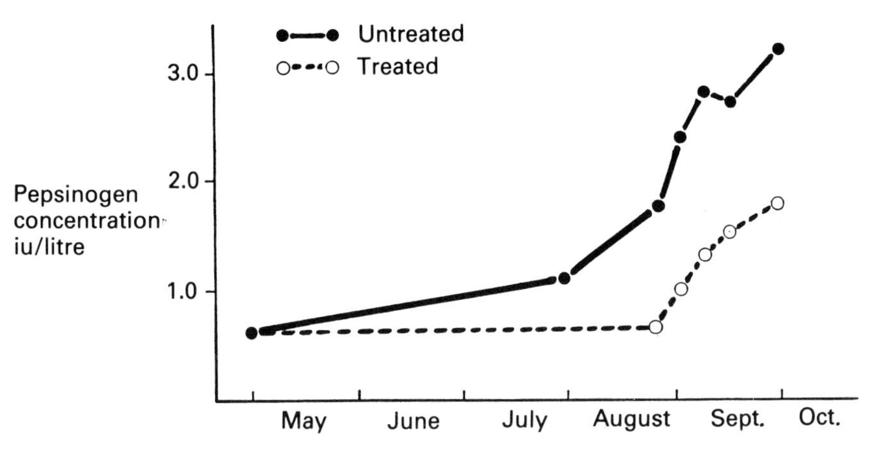

In 1987 symptoms of ill-thrift and scour necessitated the treatment of six animals with oxfendazole. In 1988, although homeopathic treatment (Santonin) was administered from mid-August, additional treatment of two animals with oxfendazole was necessary.

Parasitic gastro-enteritis is obviously an unsolved problem in this system of beef production. At present (1989), two preventive homeopathic remedies are being assessed, involving treatment prior to turnout and at regular intervals throughout the grazing season. The problem could be lessened by reducing the stocking rate or introducing a sheep flock to create a clean grazing system, but both alternatives involve restrictions in the range of management options available to the farmer.

## FINANCIAL PERFORMANCE

A financial analysis of the 1987-1989 production cycle at Aberdeen is presented in Table 5.6.

Gross output was slightly higher for I because of the greater quantity of surplus silage remaining within that system. The inclusion of surplus silage as a component of gross output may be questioned but, as indicated earlier, it is an important indicator in assessing the overall output from each of these self-contained systems.

Variable costs were higher for O than for I, primarily because of higher milk powder costs and the more costly organic cereal used in the finishing winter. Organic calves were on milk for a longer period of time, and additive-free milk powder was considerably more expensive than conventional powders. The organic oats used in this project cost £160 per tonne.

Gross margin (GM) per head was 12 percent less for O than for I but because of the lower stocking rate, the difference in GM per hectare was widened to 27 percent.

On the basis of these figures the organic system requires a premium of 30 pence per kg dead weight (i.e. 14 percent) to achieve the same gross margin per hectare after interest as the intensive system.

**Table 5.6** Financial performance of organic and intensive 18-month beef systems, 1987-89

| | O | I |
|---|---|---|
| *Output (£ per head)* | | |
| Sales at 214 p per kg | 543 | 540 |
| less calf cost | 175 | 175 |
| plus value of surplus silage at £15 per tonne | 12 | 32 |
| Output | *380* | *397* |
| *Variable costs (£ per head)* | | |
| Milk powder | 33 | 12 |
| Concentrates: | | |
| First winter | 45 | 40 |
| Grazing | 11 | 4 |
| Finishing winter | 35 | 21 |
| Forage | 8 | 35 |
| Veterinary | 2 | 9 |
| Bedding | 20 | 20 |
| Total variable costs | *154* | *141* |
| *Gross margin per head* | *226* | *256* |
| *Gross margin per hectare* | *806* | *1108* |
| Working capital per hectare | 897 | 1063 |
| Interest on capital (15% per annum) | 176 | 210 |
| *Gross margin per hectare after interest* | *630* | *898* |

## CONCLUSIONS

This organic system, based on a vigorous clover/grass sward, had a satisfactory physical and financial output, although certain technical problems remain unresolved.

Stocking rate and liveweight gain per hectare were both 18 percent less for O than for I, while gross margin per hectare after interest was 30 percent less for O, requiring a price premium of 14 percent to achieve a margin equivalent to that of I. This is not a large premium, but there may be scope for reducing even further the premium required, by minimising expensive cereal inputs through further exploitation of the nutritional benefits of clover silage.

The major technical barriers to developing a system which is sustainable in the long term are:

a) the difficulty of maintaining soil potassium status, and
b) controlling parasitic gastro-enteritis.

Improved waste handling, storage and application techniques for both manures and silage effluent will reduce leakage from the potassium cycle, but it may also be necessary to adopt a livestock system which is less dependent on ensiled herbage than 18-month beef, such as a 24-month system, although this has adverse implications for cash flow.

Similarly, with stomach worm infection a system of preventive worm control needs to be developed which is acceptable to the Standards authorities, if a viable beef system is to be sustained.

## REFERENCES

Frame, J and Newbould, P (1986) Agronomy of white clover. *Advances in Agronomy* **40**, 1-88.

MLC (1988) *Beef Yearbook 1988.* Meat and Livestock Commission, Bletchley.

Peel, S, Mayne, C S, Tichen, N M and Huckle, C A (1988) Beef production from grass/white clover swards. In *Efficient Beef Production from Grass* (Edited by J Frame). Occasional Symposium of the British Grassland Society, No **22,** 97-104.

Thomson, D J (1984) The nutritive value of white clover. In *Forage Legumes* (Edited by D J Thomson), Occasional Symposium of the British Grassland Society, No **16,** 78-92.

Younie, D, Heath, S B and Halliday, G K (1988) Factors affecting the conversion of a clover-based beef system to organic production. In *Efficient Beef Production from Grass* (Edited by J Frame), Occasional Symposium of the British Grassland Society, No **22**, 105-111.

CHAPTER 6

# ORGANIC BEEF PRODUCTION: THE PRODUCTION ECONOMICS OF ALTERNATIVE SYSTEMS

**A.W. Spedding**

Meat and Livestock Commission, PO Box 44, Winterhill House, Snowdon Drive, Milton Keynes MK6 1AX

## SUMMARY

*Results from MLC Beefplan have been used to propose production standards for similar organic systems. Premiums required for organic beef to produce similar gross margins have also been calculated.*

*Organic bucket rearing of dairy-bred calves for sale at 12 weeks requires a sale premium of £35 per calf to equal the margin for conventional calf rearing. In 18 month beef systems a premium of 49p per kg carcase weight is required on low clover swards and 32p per kg on high clover swards. Store cattle purchased in spring for grass finishing require a premium of 29p per kg carcase weight on low clover swards and 17p per kg carcase weight on high clover swards. Premiums for a lowland single suckler herd finishing its calves are even lower, at 10p and 5p per kg carcase weight for low and high clover swards respectively.*

*It is concluded that the more intensive the system of production the higher the premium needed to make organic beef margins similar to those from the conventional systems. With current production economics pushing towards higher liveweight gains, higher slaughter weights and cattle being slaughtered relatively young, organic producers need to select their system to fit their farm with care. Often the more intensive systems like 18 month beef will pay better than extensive systems like suckler herds, even when they cannot generate high enough premiums to equal margins from the conventional system.*

*Success with organic beef production will depend on running carefully planned systems with production targets tailored to fit the*

***resources available. Regular weighing and recording is vital to ensure production stays on target.***

## INTRODUCTION

The EEC ban on hormone implants in December 1986 improved the competitiveness of organic beef production *vis a vis* conventional systems, reducing the premium needed to produce similar gross margins by about 13p/kg (MLC Beef Yearbook, 1988). Thus organic beef systems should be more financially feasible than ever before.

This paper uses results recorded on commercial farms in MLC Beefplan to propose production standards for similar organic systems.

Output per hectare will be lower and some production costs will be higher for organic production than for conventional systems, and the budgets give calculations of the premiums required for organic beef to produce similar gross margins per hectare after interest has been charged on average working capital.

Practical aspects of management and husbandry of the systems described are discussed in detail in other chapters. However, all the systems budgeted have been run on commercial farms or in experimental trials.

## CALF REARING FOR SALE AT 12 WEEKS

Conventional bucket rearing usually involves feeding milk replacer once or twice daily until calves have been on the farm for about 5 weeks, after which they are eating at least 1 kg of solid food and are weaned onto concentrates and roughage. Soil Association standards require additive-free milk powder and weaning at 9 weeks. Calf rearing on the bucket will be one of the most difficult aspects of beef systems to run in practice.

Calves come from a variety of sources, possibly having been through several markets, and they arrive at their new home stressed and exposed to a whole range of new disease challenges. For this reason, conventional systems have a high dependence on modern pharmaceuticals. Nevertheless, it has been assumed in the budget that calf performance equivalent to that of conventional systems is achieved on the organic system. It has also been assumed that for organic production milk replacer and concentrates cost 10 per cent more than those for the conventional system, but that veterinary and other costs are similar.

The budget in Table 6.1 shows that a premium of £35 per calf would be necessary for the organic system to produce similar margins to the Beefplan average.

## 18 MONTH BEEF

18 month beef uses autumn-born, dairy-bred calves which are reared indoors during their first winter, grazed for one season, then finished out of yards the following spring.

Beefplan average performance and stocking rates, together with the assumptions which have been made for two different approaches to organic beef production from the 18 month system are presented in Table 6.2. It has been assumed that for organic beef:

a) cattle performance is similar to the Beefplan average;
b) feed requirements are also similar, except in the finishing winter when cutting out in-feed growth promoters would increase concentrate consumption;
c) stocking rates are halved unless clover is used;
d) clover swards have a production level equivalent to grass receiving 150 kg N fertilizer per hectare.

Financial budgets are presented in Table 6.3, with the following assumptions:

a) organic feed concentrates are 10 per cent more expensive than regular products;
b) reared calves are 17 per cent more expensive than the Beefplan average.

The analysis shows that on low clover swards a premium of 49p per kg carcase weight would be needed to produce the Beefplan average gross margin per hectare after interest charges. Using clover allows stocking rates to increase and reduces the premium required to 32p per kg carcase weight.

**Table 6.1** Beefplan average results (1987) and budgets for organic calf rearing to 3 months.

| | Beefplan average | Organic system budgets |
|---|---|---|
| *Output* (£/head) | | |
| Sales | 206 | 206 |
| Less calf cost and mortality | 133 | 146 |
| *Output* | *73* | *60* |
| *Variable costs* (£/head) | | |
| Milk replacer | 11 | 33 |
| Concentrates | 29 | 28 |
| Veterinary | 3 | 3 |
| Other costs | 3 | 3 |
| *Total variable costs* | *46* | *67* |
| GROSS MARGIN PER CALF (£/head) | 27 | −7 |
| Interest @ 15% per annum (£/head) | 5 | 6 |
| GROSS MARGIN AFTER INTEREST (£/head) | 22 | 13 |
| *Sale price required to equal Beefplan Gross Margin (£/head)* | | ***241*** |
| *Performance* | | |
| Feeding period (days) | 86 | 86 |
| Weight (kg) at start | 50 | 50 |
| at end | 116 | 116 |
| Daily gain (kg) | 0.73 | 0.73 |
| Mortality (%) | 4 | 4 |
| *Feed* | | |
| Milk replacer (kg) | 14 | 33 |
| Concentrate (kg) | 165 | 146 |

**Table 6.2.** Assumed performance levels for organic 18-month beef, together with Beefplan average herd results

| | Beefplan average | Organic system budgets | |
|---|---|---|---|
| | | Low clover | High clover |
| *Cattle performance* | | | |
| Feed period (days) | 556 | 556 | 556 |
| Daily gain (kg) | | | |
| First winter | 0.71 | 0.71 | 0.71 |
| Grazing | 0.72 | 0.72 | 0.72 |
| Finishing winter | 0.86 | 0.86 | 0.86 |
| Liveweight (kg) | | | |
| At start | 116 | 116 | 116 |
| At turnout | 199 | 199 | 199 |
| At yarding | 323 | 323 | 323 |
| At slaughter | 470 | 470 | 470 |
| Carcase weight (kg) | 254 | 254 | 254 |
| *Feeds* | | | |
| Silage (t) | | | |
| First winter | 0.9 | 0.9 | 0.9 |
| Finishing winter | 3.9 | 3.9 | 3.9 |
| Concentrates (kg) | | | |
| First winter | 130 | 130 | 130 |
| Grazing | 48 | 48 | 48 |
| Finishing winter | 473 | 550 | 550 |
| *Grazing management* | | | |
| N fertilizer (kg/hectare) | 227 | 0 | 0 |
| Stocking rate (cattle/hectare) | 3.85 | 1.90 | 3.08 |

**Table 6.3.** Budgets for organic 18-month beef production, and average results for Beefplan herds

| | Beefplan average | Organic system budgets | |
|---|---|---|---|
| | | Low clover | High clover |
| *Output* (£ per head) | | | |
| Sales @ 111p per kg lwt. | 522 | 522 | 522 |
| Less calf cost and mortality | 206 | 241 | 241 |
| *Output* | *316* | *281* | *281* |
| *Variable costs* (£ per head) | | | |
| Concentrates | | | |
| First winter | 20 | 22 | 22 |
| Grazing | 6 | 7 | 7 |
| Finishing winter | 59 | 76 | 76 |
| Forage | 36 | 10 | 30 |
| Veterinary | 12 | 12 | 12 |
| Bedding | 16 | 16 | 16 |
| Other costs | 14 | 14 | 14 |
| *Variable costs* | *163* | *157* | *177* |
| GROSS MARGIN PER HEAD (£) | 153 | 124 | 104 |
| GROSS MARGIN PER HECTARE (£) | 589 | 236 | 320 |
| Working capital per hectare (£) | 1107 | 607 | 1014 |
| Interest on capital (15% pa) (£) | 253 | 139 | 232 |
| GROSS MARGIN PER HECTARE AFTER INTEREST (£) | 336 | 97 | 88 |
| *Premium to equal Beefplan average (p/kg carcase weight)* | | ***49*** | ***32*** |

## GRASS FINISHING OF STORES

Many systems of beef production operate at a much lower intensity of grassland stocking than 18 month beef. For example, producers grass finishing stores use an average of only 1 kg N fertilizer per hectare compared to 277 kg per hectare in 18 month beef. For these producers, moving to organic beef does not involve such a fundamental change in management or production economics. A major problem, however, would be the procurement of stores guaranteed to have been reared organically.

Table 6.4 shows Beefplan performance for grass finishing and the assumptions which have been made for the alternative systems. Like 18 month beef, it has been assumed that performance is similar on all alternatives, but that stores cost more and stocking rates are lower when no N fertilizer is used.

Financial budgets for the modified systems are given with average Beefplan results in Table 6.5. The table shows that lower premiums are needed than for 18 month beef. On low clover swards a premium of 29p per kg carcase weight is needed, while on high clover swards only 17p per kg is needed to equal the Beefplan average gross margins.

**Table 6.4** Budgets for organic grass finishing of stores together with Beefplan average results

| | Beefplan average | Organic system budgets | |
|---|---|---|---|
| | | Low clover | High clover |
| *Cattle performance* | | | |
| Feeding period (days) | 144 | 144 | 144 |
| Daily gain (kg) | 0.9 | 0.9 | 0.9 |
| Weight (kg) at purchase | 348 | 348 | 348 |
| at slaughter | 479 | 479 | 479 |
| Carcase wt (kg) | 259 | 259 | 259 |
| *Feeds* | | | |
| Concentrates (kg) | 28 | 28 | 28 |
| *Grassland management* | | | |
| N fertilizer (kg/ha) | 177 | 0 | 0 |
| Stocking rate (cattle/ha) | 5.18 | 3.13 | 4.87 |

**Table 6.5** Budgets for organic grass finishing of beef stores, together with Beefplan average results

| | Beefplan average | Organic system budgets | |
|---|---|---|---|
| | | Low clover | High clover |
| *Output* (£/head) | | | |
| Sales @ 100p/kg lwt. | 479 | 479 | 479 |
| Less store cost | 368 | 405 | 405 |
| | @106p/kg lwt | @117p/kg lwt | @117p/kg lwt |
| *Output* | *111* | *74* | *74* |
| *Variable costs* (£/head) | | | |
| Concentrates | 4 | 5 | 5 |
| Forage | 12 | 8 | 12 |
| Veterinary | 4 | 4 | 4 |
| Other costs | 7 | 7 | 7 |
| *Total variable costs* | *27* | *24* | *28* |
| GROSS MARGIN PER HEAD | 84 | 50 | 46 |
| GROSS MARGIN PER HECTARE | 435 | 157 | 224 |
| Working capital per hectare (£) | 1,976 | 1,305 | 2,041 |
| Interest on working capital (15% pa) (£) | 117 | 77 | 120 |
| GROSS MARGIN PER HECTARE AFTER INTEREST | 318 | 80 | 104 |
| *Premium to equal Beefplan average (p/kg carcase weight)* | | ***29*** | ***17*** |

## SUCKLER PRODUCTION

Organic suckler systems need lower premiums to equal Beefplan gross margins than the other systems discussed. Tables 6.6 and 6.7 illustrate the reasons.

The budgets for the organic suckler systems assume a similar level of performance to that achieved by average Beefplan herds, except for stocking rates on the low clover option which would be lower. With clover, similar stocking rates to the Beefplan average should be possible. It has also been assumed that organic concentrates cost 10 per cent more than Beefplan average figures, and that herd replacement costs are 10 per cent higher.

Table 6.6 shows the Beefplan average physical performance and that which would be necessary to achieve the financial results set out in Table 6.7. The organic system without clover needs a premium of 10p per kg carcase weight to equal the Beefplan gross margin per hectare after interest on working capital.

However, the low margins for suckler production should be noted. At these levels fixed costs must be very low for a net margin to be possible. Cheap, lower quality land is necessary; on better land the dairy beef systems like 18-month beef are likely to produce better margins than suckler production.

**Table 6.6** Beefplan average results and budgets for organically run suckler herds finishing progeny overwinter

| | Beefplan average | Organic system budgets | |
|---|---|---|---|
| *Cow performance* (per 100 mated) | | | |
| Calving period (weeks) | 12 | 12 | |
| Herd replacements | 16 | 16 | |
| Herd disposals | 12 | 12 | |
| Cows barren | 6 | 6 | |
| Cows born alive | 90 | 90 | |
| Calves purchased | 4 | 4 | |
| Calf mortality | 5 | 5 | |
| *Calf performance* | | | |
| Age at yarding (days) | 372 | 372 | |
| Weight at yarding (kg) | 369 | 369 | |
| Daily gain (kg) | 0.88 | 0.88 | |
| Weight at slaughter (kg) | 460 | 460 | |
| Daily gain (finishing) (kg) | 0.9 | 0.9 | |
| Carcase weight (kg) | 258 | 258 | |
| *Feeds* | | | |
| Cow concentrate (kg) | 189 | 189 | |
| Calf concentrate (kg) | 235 | 235 | |
| Finishing concentrate (kg) | 440 | 440 | |
| Silage (tonnes) cow | 4.7 | 4.7 | |
| finishing | 3.5 | 3.5 | |
| Other feed (tonnes) | 1.0 | 1.0 | |
| *Stocking rate* (head/ha) | | *Low Clover* | *High Clover* |
| Cows | 1.66 | 1.30 | 1.66 |
| Finishing animals | 9.67 | 4.90 | 9.67 |
| Overall | 1.42 | 1.03 | 1.42 |

**Table 6.7** Beefplan average results and budgets for an organically run suckler herd finishing progeny overwinter

| | Beefplan average | Organic system budgets | |
|---|---|---|---|
| | | Low clover | High clover |
| *Output* (£/head) | | | |
| Sales | 487 | 487 | 487 |
| Cow subsidy | 31 | 31 | 31 |
| less herd replacements | 23 | 26 | 26 |
| *Output* | *495* | *492* | *492* |
| *Variable costs* (£/head) | | | |
| Cow concentrate | 22 | 24 | 24 |
| Calf creep feed | 25 | 28 | 28 |
| Finishing concentrate | 55 | 61 | 61 |
| Forage | 70 | 30 | 65 |
| Other feed | 22 | 24 | 24 |
| Veterinary | 22 | 22 | 22 |
| Bedding | 20 | 20 | 20 |
| Other costs | 18 | 18 | 18 |
| *Total variable costs* | *254* | *227* | *262* |
| GROSS MARGIN PER HEAD (£) | 241 | 265 | 230 |
| GROSS MARGIN PER HECTARE (£) | 342 | 273 | 327 |
| Average working capital (£/ha) | 1046 | 748 | 1056 |
| GROSS MARGIN PER HECTARE AFTER INTEREST (£) | 184 | 161 | 169 |
| *Increase in sale price needed to equal Beefplan margin (p/kg)* | | *10* | *5* |

## CHOOSING A SYSTEM

Results from Beefplan show that intensive systems produce higher gross margins than extensive systems.

Intensive systems are those which use high energy feeds, whether concentrates, grazing or silage, to produce finished cattle relatively young at high levels of liveweight gain and at high stocking rates. They are financially successful for conventional production because they are both biologically and financially efficient. Intensive systems are biologically more efficient than extensive systems because a higher proportion of feed goes to production and less goes to maintenance. They are financially efficient because at high liveweight gains feed cost per unit gain is lower, even allowing for the cheaper feed which can make up more of the diet for extensively kept animals. With current high interest rates, bank interest charges on working capital can be prohibitively high for extensive systems, as Table 6.8 shows.

**Table 6.8** Interest charges on average working capital for various beef systems (Based on Beefplan results, 1987)

| System | Duration (months) | Interest @ 15% per annum (£/head) |
|---|---|---|
| Calf rearing | 3 | 6 |
| Cereal beef | 12 | 39 |
| Silage beef | 16 | 44 |
| 18 month beef | 18 | 51 |
| 2 year beef | 24 | 65 |

Unfortunately for prospective producers of organic beef, intensive systems depend more than extensive systems on artificial fertilizers to allow high stocking rates and modern pharmaceuticals to control disease and internal parasites at grass. Clover can be used to make up some of the grass production lost by not using artificial products, but grass management to utilise its full potential can be tricky. The result, as these calculations show, is that the more intensive the system the higher the premium needed for organic beef.

At the farm level the system to choose will depend on the resources available. On good quality land 18 month beef may be the best system, even if the premium required to equal Beefplan cannot all be gained. On poorer land or where cheap by-products are available a suckler herd may be better.

Finally, success in beef production depends on careful management to maximise performance and stocking rates, and on painstaking marketing, with production recording playing a vital role in ensuring that performance, feed and stocking targets are met. In Beefplan there are bigger differences between top and bottom third gross margins within systems than between systems as Table 6.9 shows. As the old song says 'It ain't what you do its the way that you do it'—that is what gets results.

**Table 6.9** Top and bottom third results from various beef systems recorded by Beefplan (1987)

| | Gross margin (£/hectare) | |
|---|---|---|
| | Bottom third | Top third |
| Grass silage beef | 652 | 2002 |
| 18 month beef | 381 | 988 |
| Winter finishing | 133 | 1451 |
| Lowland sucklers | 294 | 557 |

Organic production contains all the pitfalls of conventional beef production and some extra challenges of its own. But these calculations and other evidence suggests that there should be very satisfactory profits available to efficient producers who market their produce well.

CHAPTER 7

# STANDARDS FOR ORGANIC MEAT PROCESSING AND MARKETING

**J.M. Hassett**

Billington Foods Ltd, Cunard Building, Liverpool L3 1EL

## SUMMARY

*This paper considers the new UKROFS standards as they apply to the processing and marketing of organically produced meat. The areas it covers include:*

*a) the role and structure of the UKROFS Processors Advisory Committee;*

*b) an explanation of the philosophy behind the processing standards and examples of its practical applications;*

*c) a review of the UKROFS processing standards from farm gate to supermarket shelf;*

*d) a review of the UKROFS certification scheme for meat processors;*

*e) the rules for marketing meat using the UKROFS symbol.*

## INTRODUCTION

Very early in its life the United Kingdom Register of Organic Food Standards (UKROFS) Board established two expert advisory committees to assist in the creation of its standards. These are the Producers Advisory Committee, which deals with production standards on farm, and the Processors Advisory Committee, which sets standards covering all processing activities between farm gate and supermarket shelf. The draft standards prepared by the two Advisory Committees are reviewed and approved by the UKROFS Board before publication. The work of the Producers Advisory Committee is now complete and the committee has been disbanded.

## THE PROCESSORS ADVISORY COMMITTEE

The Processors Advisory Committee continues to develop processing standards for the very wide range of single and multi-ingredient, processed organic foods for which standards have been requested.

The Chairman of the Processors Advisory Committee is a member of the UKROFS Board. Committee members are drawn from a range of processing interests in the organic movement, supported by a changing panel of co-opted members expert in the processing area under review.

The UKROFS meat processing standards have been prepared with the benefit of expert advice from the MLC, a butcher specialising in organically produced meat, and Food from Britain.

## THE PHILOSOPHY BEHIND THE PROCESSING STANDARDS

The principal purpose of the UKROFS processing standards is to protect at all times the organic integrity of UKROFS approved produce as it moves along the distribution chain from farm gate to supermarket shelf. This protection is afforded in four main ways.

The processing standards require that:

★ Organically produced produce must be transported, stored and processed separately from conventional produce so that there is no risk of jeopardising its organic integrity.

★ The records kept by the processor must provide a secure audit trail of the movement of the organic produce through his plant from receipt to final despatch.

★ There must be no break in the processing chain. All those processors who physically handle the produce and alter its nature must be registered with UKROFS for the produce to retain its UKROFS certification as it moves along the distribution chain.

★ All registered processing plants must be inspected at least once a year by an UKROFS inspector.

The following are examples of how these principles will apply in practice.

In an ideal world, organically-produced meat would be processed in a dedicated factory so that there could be no possibility of jeopardising its organic integrity through contact with conventional meat. However, the UKROFS Board recognises that at this early stage of its development the organic meat market could not commercially justify a dedicated factory, or even perhaps dedicated plant within a factory for processing organically produced meat alone. The processing standards therefore permit the use of common plant provided that organic production runs are separated from conventional production by time and approved plant cleaning procedures.

Meat will present particular problems to UKROFS inspectors checking the audit trail from farm gate to supermarket shelf. The UKROFS Board has decided that only by physically identifying organically produced meat will it be possible to trace with any hope of success its progress through abattoirs and processing plants. Accordingly, the processing standards require that all red meat carcasses be strip marked with the UKROFS symbol at the time of slaughter, and that individual carcasses carry with them throughout processing down to primal cuts their slaughter number and date. The onus of proving the audit trail through his plant will lie with the processor.

Since there must be no break in the processing chain, it follows that a processor will have to use meat grown by UKROFS registered farmers and slaughtered in UKROFS approved abattoirs for that meat to be marketed with UKROFS certification.

An UKROFS appointed inspector will visit every processing plant at least once a year. The UKROFS Board believes that regular, detailed inspection is essential to the credibility of the registration scheme. The main purpose of the inspection will be to ensure that the philosophy behind the standards has been followed in both spirit and practice.

## MEAT PROCESSING STANDARDS

In addition to a set of General Requirements, the UKROFS standards for processing organically produced meat cover welfare, specific standards for beef, pork, lamb, edible offals and poultry, and packaging materials.

The General Requirements deal with product integrity, plant hygiene, cleaning and pest control.

The standards place great importance on animal welfare, both on and off the farm. There are standards for loading on farm, transport, lairage and slaughter. If it is likely that the animals will have to be fed at the abattoir, the producer must supply feed which complies with UKROFS standards. Animals must be slaughtered humanely with a minimum of stress and in conditions that reflect proper concern for their welfare. The use of tenderising substances is prohibited.

There are beef processing standards for sides, quarters and chilled and frozen primal cuts. Like the other meat and poultry standards they cover good processing practice as well as labelling or marking to provide the audit trail. Outer packaging, where used, must carry the name and address of the supplier, details of the contents and the slaughter date.

The lamb and pork processing standards cover whole carcasses, sides and chilled and frozen primal cuts. The edible offals standards cover chilled and frozen offals.

The poultry standards cover traditional farm fresh whole body and eviscerated poultry, prepackaged fresh, and dry chilled frozen whole oven-ready poultry.

The Processors Advisory Committee plans to create other meat and poultry standards in due course, including standards for composite meat products with other ingredients such as sausages.

## CERTIFICATION SCHEME FOR MEAT PROCESSORS

Processors who wish to process and market organically-produced meat under the UKROFS umbrella must register with UKROFS. There are two ways of joining the scheme.

First, a processor may join UKROFS via an Approved Organic Sector Body. Several such organisations have already applied to join the scheme. The processor will be subject to the rules and standards of the Approved Body, including inspection, all of which will have been approved by UKROFS. The membership fee will be paid to the Approved Body and will entitle the processor to use the UKROFS symbol subject to UKROFS rules.

Alternatively, a processor may join UKROFS direct. In the first instance he should obtain a copy of the Operating Manual and the Standards from the Secretary of UKROFS. After satisfying himself that he can meet the Standards, he should complete the application form in the Operating Manual and send it to the Secretary, together with all the supporting paperwork and an application fee of £50.

An UKROFS inspector will visit the premises for which certification is sought, to determine whether the practices and procedures in the plant meet UKROFS standards. The processor will be given the opportunity to deal with reported deficiencies before an Assessment Report is sent to the Certification Committee.

The Certification Committee will review the application and make a recommendation to the UKROFS Board. If successful, the processor will be issued with a Certificate of Registration. If approval is withheld, the processor will be given reasons why and will have the opportunity to take corrective action. The Certificate of Registration can be withdrawn at any time if the Board finds that the processor is not complying with the requirements of the Certification Scheme. There is a formal appeals procedure.

It is a condition of registration that there will be periodic surveillance inspections. Their frequency will be at the Board's discretion, but there will be at least one inspection per year.

The annual cost of registration depends on the size of the organic enterprise in relation to the whole plant. There is a formula for calculating the fee based on the number of employees engaged in processing organic meat. Currently, the minimum fee is £500 per annum. There is provision in the scheme for Assessment and Surveillance Inspections to be charged at cost.

## RULES FOR USING THE UKROFS CERTIFICATION MARK

All registered processors are entitled to use the UKROFS certification mark, whether they have joined the scheme direct or through an Approved Body.

The mark may be used only on produce which has been grown

and processed to UKROFS standards. It may not be used on multi-ingredient products unless all the ingredients (with a few minor exceptions) are UKROFS approved. The mark may be used only in its approved form, as set out in the Manual.

If a registered processor does not wish to use the certification mark, perhaps because he prefers to use the symbol of an Approved Organic Sector Body, then there is an agreed form of wording to denote UKROFS approval which may be used instead.

## REVIEW OF PROGRESS

This paper has sought to explain the background to the UKROFS scheme for the processing and marketing of organic meat. The standards will be subject to regular review in the light of practical experience, and the Processors Advisory Committee will welcome advice from registered processors once they have that experience.

After two years the UKROFS Board believes it has put in place a viable and credible scheme which will enable producers and processors to enter with confidence the burgeoning market for organic meat.

CHAPTER 8

# ORGANIC MEAT MARKETS: A MAJOR FOOD RETAILER'S APPROACH

M. Hunt
Safeway Plc, Argyll House, Millington Road, Hayes, Middlesex

## SUMMARY

*Multiple grocers have increased their meat sales by 24 per cent in the last 5 years, while traditional butchers have lost 18 per cent of their market share, and this trend will continue. Several social and economic changes have taken place over the last few years which have affected consumer attitudes to the purchase of food. Trends indicate that the convenience factor, which is dominant at present, will be superseded by a requirement for fresh food, free from additives and organically farmed where possible. Meat has received adverse publicity healthwise, but most people enjoy eating meat and purchases must be encouraged. Safeway has strict buying policies and quality control requiring large numbers of animals within a specific weight range, well proportioned and with a suitable level of fat. Safeway would like to include organic meat in its range of organic products, but there are three main problems to be overcome: there must be continuity of supply of meat of a guaranteed quality which can be sold profitably at an acceptable price. The production of organic meat requires a premium for the producer to achieve similar gross margins to those of a conventional system. However, the retailer also incurs additional costs in segregating carcasses from conventional meat in the processing plant, and in selling surplus parts of the carcase as normal meat. Costs must be curtailed as far as possible to allow organic meat to be sold at an acceptable price.*

## INTRODUCTION

As a retailer, I firmly believe in the necessity for each of us to understand the relative position of our various sectors within the meat industry and to further the cause of total communication.

between those sectors. My company is a major multiple retailer of all forms of food, not just red meat.

Safeway plc has developed over the years through the acquisition of many companies, some of which were founded in the last century. In 1986 the group acquired the UK division of Safeway from its American parent and this comprised 112 highly successful stores. After looking closely at the constituent parts of the group as it then stood, trading under the three different facias of Safeway, Presto and Lo-Cost, we decided that a new approach was needed. Presto has always consisted of a wide range of stores from virtually hypermarket size to convenience store level; Lo-Cost consists of mainly small units whereas Safeway stores are relatively large. Of these facias it was an accepted fact that Safeway was the name that had the highest profile nationally with the consumer.

It was then decided that we would embark on a major conversion programme in which all the large Prestos became Safeway and the smaller ones converted to Lo-Cost. This programme is currently in full swing. In 1986 there were 112 Safeway stores, and by the end of 1988 this became 220. It is now 268 and the target is 450 by 1991. Included in these figures are new stores which we are opening at a rate of 23 each year.

We employ 65 thousand people and have a selling area of 7.3 million square feet; there are twelve distribution centres but only one buying centre. The total sales for the Argyll Group, for the year ending March 1989, was £3.7 billion, and we are currently joint third in the league.

## PURCHASE OF MEAT

There are major changes taking place in the source of meat purchases. Traditional butchers have lost 18 per cent of their market share within the last five years with multiple grocers increasing theirs by 24 per cent.

This has and will continue as the high street butcher is driven out by increasing costs and falling trade, to the point where he is relegated to suburban and rural sites. The butchers that will survive are the specialists and the experts who can provide skills and services that no multiple retailer can compete with.

The supermarkets' share will continue to increase, providing the one stop convenience shopping that the consumer increasingly demands.

## THE CONSUMER

Over the last few years many changes have taken place, socially and economically, that have affected both the supply and sales of fresh meat. Consumer attitudes and requirements have dramatically altered, reflecting changing lifestyles and increasing awareness of the product they buy. Working habits, particularly among women, have had a considerable influence on food requirements. Convenience is now the prime consideration, hence the rapid growth in the ready meals and further processed sector.

Attitudes to food preparation have changed, especially with the younger generation who are no longer trained or experienced in the traditional handling of raw foods. This has had a detrimental effect on several markets such as offal, fresh fish and, at the cheaper end of the red meats range, cuts such as brisket of beef or breast of lamb are no longer popular; modern youth shows preference for a meal of fish fingers or burgers and chips. What will be their requirement when they become the shoppers of the future?

Increased leisure time and facilities have affected the Sunday lunch as the main meal occasion of the week. The demand for joints of beef, pork or lamb is declining as a traditional weekly purchase, becoming more and more a special occasion meal. Once again this has increased the demand for convenience both in purchase and preparation.

One positive change has been the increased range of foreign foods being consumed. Holidays abroad have become the norm. Eating experience gained in other countries, coupled with the proliferation of ethnic restaurants and take-aways within our towns and cities, has led to the consumer experimenting with foreign dishes at home, either self-prepared or pre-packed.

There is a growing awareness and concern among consumers about what they eat and what it contains. Additives are now taboo with a large sector of the public, and the statement 'additive free' has become a major selling point.

Fortunately, recent moves within the farming industry on the use of hormone implants and more control over the use of antibiotics allows us increasingly to claim that meat is a natural food. Trends indicate that, although the convenience factor is dominant at the moment, this will be superseded by a requirement for fresh food, free from additives, organically farmed where possible, with the overall impression of being natural and healthy.

## HEALTH ASPECTS OF MEAT

There is a major adversity that affects the meat industry both in its conventional and organic form, namely '*meat kills*'. Over the last few years there has been a wealth of medical opinion, for example the COMA report, on the dangers on fat within the human diet, with red meat being particularly targeted as a major factor in the increasing incidence of heart disease. Although the meat industry has fought against some of the accusations, which are inaccurate, biased and in some cases totally untrue, the main headlines in the media still focus on meat eating being unhealthy with the positive points if mentioned at all being relegated to the small print.

The basic fact is that excess fat on meat is bad for you. Lean red meat is as low in fat as chicken, usually perceived as healthy food, and is a positive and even necessary part of a healthy and nutritious diet. The most positive point of all is that the majority of the population actually like and enjoy eating meat.

## MEAT RETAILING BY SAFEWAY

What meat does Safeway buy, how do we buy it and where do we buy from?

This year Safeway will buy beef, pork and lamb to a value of approximately £150 million pounds. In round figures we will require:

90,000 cattle, 380,000 pigs and 400,000 lambs.

This meat will be purchased through abattoirs that have passed Safeway's stringent quality control inspections and which would more than likely be EEC approved. It will be purchased against strict specifications:

*Beef* to MLC classifications R, U or E for conformation, and 2, 3 or 4L for fat cover. The carcase weight will be between 240 and 360 kg.

*Lamb* will be classified R, U or E for conformation and 2 to 3L for fat cover. Weight will range from 16 to 20 kg.

*Pork* would be expected to have a probe measurement of between 18 and 24 and a head-on weight range of 55 to 73 kg.

Basically we need an animal within a specific weight range, which is well-proportioned, neither too fat nor too lean, and which will give us both a profitable yield and retail cuts to suit consumer requirements.

We enforce strict controls on pre-slaughter care, the slaughter process and post-slaughter handling, especially on the chilling of the carcase. Beef for instance will not be chilled below 10°C within the first 12 hours; it is then reduced to a deep muscle temperature of 4°C or below within the following 36 to 48 hours prior to delivery.

We have two dedicated beef processing plants where we take carcases direct from the abattoir, cut and pack them into primals and produce our fresh mince requirements into retail packs. We take between 60 percent and 70 percent of our beef through these plants with the balance of our requirement made up of purchases from approved boning plants. This system ensures that we have total control of the majority of beef destined for our stores, from the individual inspection of every carcase on receipt through to the end product.

We have a central distribution system comprising three (soon to be five) geographically placed chilled depots through which all fresh meat is delivered to the stores. Beef, pork and lamb are supplied to the stores as primals in cases from which our in-store meat departments process their retail packs.

We cut and pack approximately 70 percent of our requirement at store level, with the remaining 30 percent comprising supplier-processed fresh or frozen retail packs.

Although this only gives a basic overall view of our systems it demonstrates the scale of our operation and the volumes involved.

## ORGANIC MEAT

Firstly, I would like to emphasise Safeway's desire to include organic meat in the range that we offer our customers. The commitment is there and detailed research is currently being carried out to achieve this aim.

Safeway has been the leader in retailing and developing organic produce, which has given us an insight into the problems involved. Initially the main problem was lack of supply and continuity, which restricted the number of stores in which we could offer the product. Most growers were very dedicated but also very inexperienced, lacking the skills both to produce a product of acceptable quality and to market that product commercially. To overcome the variable supply and quality problem we have encouraged the formation of cooperatives, which has proved highly successful.

We have also been forced to seek products from abroad, partly through the seasonality inherent in the production of the products but also to ensure continuity of supply. Annually between 60 percent and 70 percent of our requirements is imported.

It seems unfair that countries such as Germany have been encouraged and supported by their governments to increase the production of organic food, whereas in the UK development has been left to the limited resources of the dedicated few.

We are actively encouraging the expansion of organic fruit and vegetables within the UK, especially with our large commercial growers, by demonstrating both the growth to date and potential we see in the future.

From the experience gained in 8 years of retailing organic produce we have ascertained guidelines that we must achieve if *we* are to expand the availability of organic meat to the British consumer.

To date only a few specialist butchers and farm shops have offered organic meat to the general public, reflecting the limited supply. For Safeway or any other major multiple to launch successfully and expand the sale of organic meat we must have the following:

★ Guaranteed volume
★ Guaranteed quality
★ Acceptable price levels

## Guaranteed volume

Safeway would like to introduce organic meat. We appreciate that the number of stores would be restricted by supply, so it could be any number from one upwards. But to launch the product successfully, the store or stores would have to offer organic meat 100 percent of the time.

To date, several farmers have approached us with the offer of organic cattle, one tomorrow, one in two months time and two for Christmas. This type of supply situation obviously does not lend itself to successful marketing within the multiple retail sector.

Let us take an example of 10 cattle per week. To fit into our distribution system these cattle would need to be processed through one abattoir, which should be Safeway approved. This creates the problem of where the cattle come from. It is unlikely that one farmer could supply this number every week of the year, so therefore we would need producer groups to be set up in conjunction with a central slaughter point. This would give the additional problem that the producers and that abattoir would have to be geographically placed to ensure that the transport of the livestock was both practical and humane.

This is a simple example but illustrates the need for an organised system to be set up.

## Guaranteed quality

As previously stated, Safeway have a carcase specification for weight conformation and fat cover which ensures that the meat we purchase is commercially viable and gives the end product that the consumer requires. Any move into organic meat would still have this specification imposed, albeit with some flexibility. We would not be prepared to accept any livestock, just because it was organically produced.

One of the major problems when we first moved into organic fruit and vegetables was that most producers were amateurs, growing on third rate land and producing a third rate product. This is not the way to go with meat. If we are to progress to any reasonable volume, farmers must look to the market place and achieve the standards required.

### Acceptable price level

What is the consumer prepared to pay for organic meat?

Already a very small sector are buying organic meat at anything up to double the price of conventional meat, but it must be said that the majority of these consumers have incomes to match their principles. For organic meat to widen its market an acceptable price level must be achieved, so that it is affordable to most consumers. What is an acceptable level has yet to be ascertained but it certainly will not be twice the normal. This means that to expand the market, organic meat must be produced, processed and retailed as economically as possible.

It is accepted that there are additional costs at farm level entailed in the production of organic livestock, but these costs must be controlled by the selection of the right breeding stock, choice of the right land and the maintenance of good husbandry.

At the abattoir additional costs are incurred in the segregation of the livestock and carcases from conventional meat. These costs can obviously be reduced as volumes increase.

Finally, at retail level there are many factors that will affect the end price.

We would not initially stop selling conventional meat within any of our stores but would retail organic meat, as we do other organic produce, as an additional choice to the consumer. This means that to ensure correct segregation from the conventional meat we would have all retail packs of organic meat produced away from the stores by approved processors. The ideal method of packaging would be controlled atmosphere packs, but we may have to look at other forms of packaging to maintain an environmentally friendly image. The method of packaging could obviously create additional cost.

One of the major factors affecting the retail price of any cut of meat is the imbalance of consumer requirements.

When a carrot is grown it is grown as a carrot, processed as a carrot, and sold as a carrot. It is not as easy with meat. We have to sell the whole animal divided into a multitude of cuts, each with an individual level of demand.

In general terms multiple retailers require two hind quarters to every fore quarter, although this will vary dependent on the time of year; this means that we actually require three legged cattle. Looking at our own requirement in Safeway, in one week we would require for example the topsides from 1700 cattle, the rumps from 1800 cattle, the standing ribs from 800 cattle and the briskets from 1200 cattle.

The days are gone when the butcher bought, killed and sold the entire animal. The meat industry is now geared to match these imbalances through the various sectors of manufacturing, catering and assorted types of retail outlet. Each part of the animal has its individual value decided by its position within the mix.

Organic meat will not have the advantage of this established market. This could mean that out of every beast produced 25 percent of the meat would have to be absorbed into the conventional market at conventional prices, thus increasing the cost of the remaining 75 percent.

It has been mooted that commercially grown, organic cattle would require a 15 percent premium, so if we assumed that all additional costs equated to conventional meat this 15 percent could then be related through to retail prices, which would mean that gross margin percentages would fall but net cash return would remain the same as conventional meat.

If possible additional costs are taken into account, the 15 percent could become 20 percent allowing for the imbalance factor. The 20 percent could become 30 percent with additional abattoir processing and packaging costs.

If yields of saleable meat from the carcase fell, by the inability of organic growers to produce livestock to conventional levels, this 30 percent could become 40 percent.

Finally, if all sectors required to maintain gross margin percentages the 40 percent becomes 50 percent. This would therefore produce retail prices such as:

★ Sirloin steak, currently just under £5 per lb, costing around £7.50 per lb.

★ Topside, currently around £2.60 per lb, becoming around £4 per lb.

★ Braising steak, at around £2 per lb, becoming £3 per lb.

I believe that this type of price structure would be totally unacceptable, restricting the appeal to a minority of consumers. It is basically up to all of us to ensure that additional costs are kept to the minimum, enabling a realistic retail pricing policy to be achieved.

## REQUIREMENTS OF A MAJOR NATIONAL MULTIPLE

If Safeway is to move into the organic meat market:

★ Farmers must form themselves into producer groups, preferably in conjunction with major EEC approved abattoirs, to ensure a guaranteed volume 52 weeks of the year.

★ These groups must ensure that not only is the product guaranteed organic within the current guidelines of either UKROFS or the Soil Association, but it is also of the right quality to compete against conventional meat.

★ Finally, costs must be controlled to ensure that the end price is conducive to market growth.

## ORGANIC MEAT IN PERSPECTIVE

Volumes are not there at this moment in time, and any growth in volume will be long term. It will never be a major competitor to conventional meat, but if it is to create a reasonable niche in the market place some of the current attitudes will have to change.

The April-June issue of the "Living Earth", in its campaign for safe meat lists only 15 retail outlets nationally that sell approved organic meat, which reflects both the lack of availability and the current low level of sales.

At the same time, publicity sought using what can only be described as fear tactics, highlighted by the cover of this magazine, is totally non-productive. Until a viable alternative is available the only result achieved is to stop people eating meat altogether, which would affect future sales of organic meat. This is a very short sighted view as over 2.2 million tonnes of beef, pork and lamb will be produced in 1989 and in 1988 the consumer spend on beef, pork and lamb was nearly £1.5 billion pounds; a 5 percent share of that market would equate to £75 million.

The crusading approach must be taken out of organic meat. It must become a strictly commercial venture. The history of the world has revolved around commerce and human nature being what it is, it always will, so my message to organic meat producers is:

*Organise, commercialise* and *professionalise* if you wish to expand your market.

## THE FUTURE FOR SAFEWAY

What will Safeway do to encourage the growth of the organic meat sector?

As I have previously stated, we wish to sell organic meat and we will give every assistance we can to develop the market. We do and will participate in research, and we are prepared to advise and assist any producer groups that approach us with a serious commercial proposition. We are even prepared to look at contracts to enable planned production to be carried out.

We will help, but essentially *action* is needed by organic meat producers *themselves.*